PLANT
PROTECTION

Managing Greenhouse Insect and Mite Pests

PLANT PROTECTION

Managing Greenhouse Insect and Mite Pests

Raymond A. Cloyd

Ball Publishing   Batavia, Illinois

Ball Publishing
335 N. River Street
P.O. Box 9
Batavia, IL 60510
www.ballpublishing.com

Copyright © 2007 by Raymond A. Cloyd. All rights reserved.

Edited by Jayne VanderVelde.
Cover and interior designed by Anissa Lobrillo.

No part of this book may be reproduced or transmitted in any form or by any means, electronic or mechanical, including photocopying, recording, or any information storage-and-retrieval system, without permission in writing from the publisher.

Disclaimer of liabilities: Reference in the publication to a trademark, proprietary product, or company name is intended for explicit description only and does not imply approval or recommendation to the exclusion of others that may be suitable.

While every effort has been made to ensure the accuracy and effectiveness of the information in this book, Ball Publishing makes no guarantee, express or implied, as to the procedures contained herein. Neither the author nor the publisher will be liable for direct, indirect, incidental, or consequential damages in connection with or arising from the furnishing, performance, or use of this book.

Library of Congress Cataloging-in-Publication Data

Cloyd, Raymond A.
  Plant protection : managing greenhouse insect and mite pests / Raymond A. Cloyd.
      p. cm.
   Includes bibliographical references.
   ISBN 978-1-883052-60-7 (softcover : alk. paper)
  1. Greenhouse plants--Diseases and pests--Control. 2. Insect pests--Identification. 3. Mites--Identification. 4. Insecticides. 5. Acaricides. I. Title.

SB608.G82C56 2007
635'.0483--dc22
                              2007014809

ISBN: 978-1-883052-60-7

Printed in Singapore by Imago.

13 12 11 10 09 08 07   1 2 3 4 5 6 7 8 9

Contents

Introduction VII

Chapter 1: Insect and Mite Feeding Behaviors 1

Chapter 2: Effects of pH on Insecticides and Miticides 9

Chapter 3: Mode of Action and Rotation 13

Chapter 4: Tank Mixing 29

Chapter 5: Coverage and Timing of Application 33

Chapter 6: Plant Phytotoxicity 41

Chapter 7: Pesticide Storage 49

Chapter 8: Identification of Insect and Mite Pests 55

About the Author 81

Selected References 83

Index 85

Introduction

Pest management in greenhouses is important in order to minimize problems with insect and mite pests as well as to produce quality crops. There are a variety of management strategies that either avoid or minimize dealing with plant-feeding insect and mite pests, including cultural, physical, and biological solutions and the use of insecticides and miticides. Cultural management involves those practices related to maintaining plant health, such as irrigation and fertility, and sanitation, which includes weed, plant, and growing medium debris removal. Physical management is associated with impeding the ability of insects to migrate into greenhouses by using screening materials. Biological management involves the use of living organisms or natural enemies, such as parasitoids, predators, and pathogens, to regulate insect and mite pests in greenhouses.

The focus of this publication will be on the proper use of insecticides and miticides in greenhouses because insecticides and miticides are still the primary means of dealing with insect and mite pests. However, greenhouse producers should not rely solely on insecticides and miticides; instead, they should consider implementing all the management strategies described above simultaneously in order to reduce insect and mite pest outbreaks, which may, in fact, lessen the need for insecticide or miticide applications. Another very important management procedure is scouting on a regular basis in order to detect insect or mite pest populations before they reach damaging levels, thus enhancing the effectiveness of insecticide or miticide applications.

First, it is critical to properly identify a given pest or pests in order to determine whether you are dealing with an insect or mite problem. This is especially important since many of the newer insecticides and miticides are selective in the range of insects or mites that they will control. For example, some insecticides or miticides only control one group of pests (i.e., mites), whereas others may control two to three different types of insects or mites. In order to properly identify a given pest or pests, greenhouse producers should have a number of reference guides or publications that contain clearly visible photographs so that it is possible to identify the pest or pests in question. (Refer to Selected References for a listing.) Another viable option is to send samples to a state extension entomologist or state plant diagnostic clinic. Once a pest or pests have been correctly identified, then the appropriate insecticide or miticide can be selected and applied.

This publication will discuss, in detail, a variety of topics including insect and mite feeding behaviors, the effects of pH on insecticides and miticides, mode of action and rotation, tank-mixing, coverage and timing of application, plant phytotoxicity, and pesticide storage. In addition, chapter 8 describes the biology, damage, and management guidelines for common insect and mite pests of greenhouses. Understanding these topics will enhance your ability to deal with plant-feeding insects and mites using insecticides or miticides.

Happy bugging!
RAC

Chapter 1

Insect and Mite Feeding Behaviors

Insects and mites that feed on plants have different feeding behaviors, which include chewing, piercing-sucking, mining, boring, or galling. The majority of insect and mite pests that attack horticultural crops grown in greenhouses have either piercing-sucking or chewing mouthparts. Insects with piercing-sucking mouthparts include aphids, whiteflies, mealybugs, and scales. These insects insert their mouthparts into the vascular tissues of plants, primarily in the food-conducting tissues (phloem), and withdraw plant fluids. This results in characteristic symptoms such as plant wilting, stunting, and leaf distortion. Insects with chewing mouthparts include beetles, caterpillars, weevils, and fungus gnat larvae (although not considered insects, snails and slugs also have chewing mouthparts). Chewing insects physically remove portions of leaves, flowers, or roots directly, or they consume entire plant parts.

The reproductive portions of plants are typically fed upon by most plant-feeding insects because these portions are more nutritious due to the relatively high concentrations of protein. However, these plant portions may also contain high concentrations of secondary metabolites (defensive compounds), which may influence acceptability. Plant leaves usually provide the greatest biomass for insects and are the best food nutritionally—next to reproductive portions. Nitrogen is the primary plant nutrient required by insects with piercing-sucking and chewing mouthparts. Almost 90 percent of the nitrogen in plants is present in the form of free proteins and amino acids, which are very important for insect survival, growth, development, and reproduction. Nitrogen concentrations are usually higher in younger tissue than in older leaves, and these concentrations decline as plants mature. The concentration of nitrogen is a factor that can limit insect

Figure 1.1
Oleander aphid (*Aphis nerii*) uses its piercing-sucking mouthparts to feed within the food-conducting tissues (phloem) of plants.

growth and development. In general, plants tend to lack nitrogen in the form that insects can utilize. The dry weight of most insects is between 8 and 14 percent nitrogen; however, plants overall only contain 2 to 4 percent nitrogen. The phloem typically contains 0.5 percent or less and the xylem (water-conducting tissues) contains 0.1 percent or less nitrogen. Reproductive portions (flowers and seeds) and leaves contain from 1 to 5 percent or more nitrogen. Protein is an important component of nitrogen; however, protein concentrations can vary depending on plant type, age, and nutritional status of the growing medium. Protein is generally higher in reproductive portions as well as leaves and stems.

Plants that are overfertilized, especially with nitrogen-based fertilizers containing ammonium, urea, and nitrate, tend to produce succulent growth, which increases their susceptibility to plant-feeding insects and mites. The higher levels of amino acids, which are the primary food source used by insects and mites, can increase their reproductive ability. In addition, plants that receive excessive levels of fertilizer may have thinner leaf cuticles or leaf laminas, which are easier for insects and mites to penetrate with their mouthparts. The survival, development rate, and reproductive capacity of insect pests with piercing-sucking mouthparts are influenced by changes in the nitrogen concentration in plants, which is typically associated with fer-

Figure 1.2

Leafminer larvae damage on chrysanthemum (*Dendranthema* x *grandiflorum*)

tility. Aphids, in particular, respond positively to increased nitrogen concentrations in host plants as a result of excessive fertilizer applications.

Variegated plants—those with white, yellow, or red coloration along with green portions—are typically fed upon more by insects because the variegated areas contain more nutrients and fewer defensive compounds than the green portions. Additionally, variegated portions are softer and easier for insects to penetrate with their piercing-sucking mouthparts.

Insects with piercing-sucking mouthparts that feed in the phloem may produce large quantities of honeydew, which is a clear, sticky liquid. Free amino acids, which are essential in the production of protein, are very important to phloem-feeding insects. These insects require protein (in the form of free amino acids) for development and reproduction. Insects with piercing-sucking mouthparts can absorb and utilize free amino acids directly from plants. These free amino acids and amides are the major sources of dietary nitrogen for aphids, whiteflies, and mealybugs. However, in order to obtain the necessary quantities of amino acids, insects must ingest large amounts of plant sap, which contains an assortment of other materials in larger quantities than amino acids, including sugars. The excess is then excreted as honeydew. In addition, phloem-feeding insects possess carbohydrases, such as amylase, and a pectin-hyrolysing enzyme that break down the middle lamellae of

4 Plant Protection

Figure 1.3
Twospotted spider mites (*Tetranychus urticae*), such as the adult shown here, feed within the spongy mesophyll, palisade parenchyma, and chloroplasts.

plant cell walls. Insects that feed within the phloem, such as some soft scales and certain plant bugs, tend to exhibit a high degree of host specificity because certain plant-specific chemical compounds tend to serve as important host selection cues whereas other insects, including aphids, mealybugs, and whiteflies, feed on a wide range of plant types. Although some aphids, including green peach (*Myzus persicae*) and melon (*Aphis gossypii*), may prefer certain cultivars of chrysanthemum (*Dendranthema grandiflora*) to others, in general they will feed on most cultivars.

Insects that feed in the xylem, such as true bugs (Order: Hemiptera), must cope with negative tension and very low concentrations of nutrients in the xylem fluid. As a result, these insects feed faster as the water potential becomes more negative and they extract extremely large quantities of plant

Figure 1.4

Adult flea beetles possess chewing mouthparts.

fluids, which is one reason why xylem-feeders tend to be larger than phloem-feeders.

Insects with piercing-sucking mouthparts that feed within the food-conducting tissues (figure 1.1) are susceptible to applications of systemic insecticides. Systemic insecticides are applied to the leaves, stem, or growing medium. Systemic insecticides are generally water soluble (expressed as either grams per liter [g/L] or parts per million [ppm]), which allows them to be distributed into roots and leaves. Also, plants do not readily metabolize systemic insecticides. In general, the active ingredient is taken up and moved throughout the plant (translocated) in the water-conducting or food-conducting tissues, or both. Additionally, once inside the plant, the active ingredient may move back and forth from the water-conducting tissues to the food-conduct-

Figure 1.5

Fungus gnat (*Bradysia* spp.) larvae have chewing mouthparts, which they use to feed on plant roots and tunnel into cuttings or plant crowns, such as the inside of this poinsettia (*Euphorbia pulcherrima*) cutting.

ing tissues or vice versa; however, this depends on the physical and molecular properties of the systemic insecticide. As an insect feeds, it takes up a lethal dose of the insecticide and is killed. For example, the piercing-sucking mouthpart or proboscis of an aphid is inserted into plant tissue and reaches the conductive cells (i.e., sieve tubes) through which water and food are transported. The aphid takes up the insecticide as it withdraws fluids.

Most of the currently available systemic insecticides are associated with two modes of action: nicotinic acetylcholine receptor disruptors and selective feeding blockers. The neonicotinoid-based insecticides kill target insect pests in a manner similar to the natural product nicotine by acting on the central nervous system, causing irreversible blockage of the postsynaptic nicotinergic acetylcholine receptors. These systemic insecticides disrupt nerve transmission in insects, causing uncontrolled firing of nerves. This results in rapid pulses from the steady influx of sodium, leading to hyperexitation, convulsions, paralysis, and death. A concern associated with using systemic insecticides having a single or site-specific mode of activity, such as the neonicotinoids, is that the selection pressure placed on insect pests from continual use of these systemic insecticides may result in the development of resistant genotypes or biotypes.

Several insecticides are classified as selective feeding blockers, which have a broad or physical mode of activity. These products kill aphids and whiteflies by blocking their stylet (feeding tube), which prevents them from feeding. As a result, the insects starve to death. This mode of action is less susceptible to insects developing resistance—in the short term. However, continued use of this mode of action for long periods of time may eventually reduce the effectiveness of these systemic insecticides.

Insects and mites with piercing-sucking mouthparts that feed primarily on the underside of leaves, such as whiteflies and twospotted spider mites, are susceptible to insecticides and miticides with translaminar properties or local systemic activity. After an application, these materials penetrate leaf tissues and form a reservoir of active ingredient within the leaf. This provides residual activity against foliar-feeding insects and mites. There are a number of insecticides and miticides currently available with translaminar activity.

Western flower thrips (*Frankliniella occidentalis*) also have piercing-sucking mouthparts; however, they tend to pierce holes in the cell walls of leaf tissue using a single stylet in the mandible and then insert a set of paired stylets, which withdraw plant fluids. As a result, they can feed on many food types within plants. Because western flower thrips do not feed extensively in the phloem tissues, systemic insecticides are less effective in controlling them. Leafminer larvae feed between the leaf surfaces in the mesophyll layer of cells (figure 1.2). Generally, this protects the larvae from applications of contact insecticides; however, products with translaminar properties are effective against the larvae as these materials are capable of entering the leaf and killing the larvae.

Spider mites—including twospotted spider mites, Lewis mites, broad mites, cyclamen mites, and eriophyid mites—do not feed in the vascular tissues. Twospotted spider mites (*Tetranychus urticae*) (figure 1.3) feed primarily on leaf undersides with their stylet-like mouthparts, damaging the spongy mesophyll, palisade parenchyma, and chloroplasts, which reduces chlorophyll content and the plant's ability to manufacture food via photosynthesis. Because mites do not feed in the vascular tissues, in general they are not susceptible to systemic insecticides; however, mites, depending on the species, are more susceptible to insecticides and miticides with translaminar properties.

Chewing insects (figures 1.4 and 1.5), in general, are non-selective in their feeding behaviors—they typically ingest macerated whole leaf or root tissue. However, some are more selective. For example, certain caterpillar leafminers consume the inner tissues of leaves but do not feed upon the less palatable outer whole tissue. Specific chewing insects, such as caterpillars, prefer food with high water content and will select plant leaves with the highest concentrations of water. Systemic insecticides are generally less effective in controlling chewing insects. Insecticides with contact and stomach-poison activity are more effective in controlling these insects. However, some systemic insecticides are effective via contact and ingestion. For example, the neonicotinoid-based insecticides will control leaf-feeding beetles and certain leafminers.

Insecticides and miticides may be applied as preventative or curative treatments depending on the insect or mite pest or pests. For example, systemic insecticides, when applied to the growing medium, need to be applied preventatively in order to control phloem-feeding insects, such as whiteflies, aphids, and mealybugs. If systemic insecticides are applied after insect pest populations are already established on plants or after plants have developed woody tissue, this may delay control; thus, insect pests may still cause damage before ingesting enough active ingredient to kill them. It is important to note that systemic insecticides must be applied when plants have an extensive, well-established root system and are actively growing in order to enhance the uptake of the active ingredient throughout the plant. Most contact insecticides and miticides are applied as curative treatments (i.e., when insect and mite pests are present). However, applications of contact insecticides or miticides need to be performed before pest populations are excessive, since any dead insect or mite pests may still adhere to plants possibly impacting acceptance and salability.

Chapter 2

Effects of pH on Insecticides and Miticides

Resistance is often blamed for the failure of an insecticide or miticide to manage insect or mite pest populations. However, inadequate insecticide or miticide performance may be associated with the pH of the spray solution. Greenhouse producers generally monitor the pH levels of water and the growing medium to maintain plant health. However, monitoring water pH can also help to maintain the effectiveness of many insecticides and miticides used to manage insect and mite pests in greenhouses.

pH is a measurement of the concentration of hydrogen ions [H^+] in a solution. It is a logarithmic scale indicating the acidic and basic properties of water. The pH scale ranges from 0 to 14. A pH value below 7 is acidic, whereas a pH value above 7 is basic or alkaline. A pH of 7 is neutral. Many common insecticides and miticides are susceptible to fragmentation if the pH of the water is not within an acceptable range. When the pH is greater than 7, then a process known as alkaline hydrolysis occurs. Alkaline hydrolysis is a degradation process in which the alkaline water breaks apart insecticide or miticide molecules. This process releases individual ions (electrically charged atoms), which may then reassemble with other ions. These new combinations may not have any insecticidal or miticidal properties.

Insecticides and miticides are more susceptible to alkaline hydrolysis than fungicides. Many insecticides and miticides are very sensitive to alkaline conditions—some in fact will degrade within a few hours after dilution in alkaline water. The carbamate and organophosphate chemical classes are generally more susceptible than pyrethroids. However, high pH levels can affect other insecticides. For example, a pH above 8 can reduce the efficacy of the *Bacillus thuringiensis* toxin and the insect growth regulator azadirachtin. Table 2.1

TABLE 2.1.
Effect of pray solution pH on the half-life* of carbaryl

SPRAY SOLUTION PH	CARBARYL HALF-LIFE
3.0	Stable (>350 weeks)
4.5	Stable (300 weeks)
5.0	Stable (100 weeks)
6.0	Stable (58 weeks)
7.0	10.5 days
8.0	1.3 days
9.0	2.5 hours
10.0	15 minutes

* Half-life is the time it takes for an insecticide to lose half of its original toxicity and effectiveness.

demonstrates the effects of pH on the insecticide carbaryl. Although this insecticide is not registered for use inside greenhouses, it demonstrates that water pH can significantly affect insecticide half-life, which is the time it takes for 50 percent of an insecticide to break down.

High temperatures increase the rate of insecticide degradation because alkaline hydrolysis occurs more rapidly under extreme temperatures. For example, at a pH of 9 and a water temperature of 77° F (24.7° C), acephate loses 50 percent of its activity in approximately two to three days and fenvalerate loses 50 percent of its activity in one to two days. In fact, doubling the temperature will double the speed of insecticide degradation. The ways to avoid water pH problems include:

1) Follow manufacturer directions on the desired water pH. The ideal pH range for most insecticides and miticides is 5.0 to 6.5. Below are examples from manufacturer labels:
    ◎ Azadirachtin will break down in spray tank mixtures that have pH values exceeding 7. The recommended pH range is 5.5 to 6.5; adjust spray solution to between 3 to 7 pH, if necessary.

- Bifenazate solutions must be used promptly to prevent degradation under alkaline conditions. Alternatively, a commercially available buffering adjuvant can be added to the solution to reduce the pH to a neutral/acidic range.

- Methiocarb performance may be reduced when spray solution has a pH greater than 7. The pH of the spray solution must be corrected by the addition of a suitable buffering or acidifying agent for optimum activity.

2) Test the water pH on a regular basis because the pH of water can change during the season. Water samples should be collected in a clean, non-reactive container, such as a glass jar. Be sure to sample water that is representative of the spray solution. Let the water run long enough so that water standing in the hose and pipes is "flushed out." Determine the pH of the water solution immediately after collection. Use an electronic meter to measure the pH.

3) Apply insecticides and miticides as soon as possible after mixing. It is advisable to use a spray mixture within six hours or less to avoid potential pH problems.

4) Do not leave insecticide or miticide solutions in a spray tank for an extended period of time (i.e., more than three hours).

5) Adjust water pH with buffers or water conditioning agents. Buffers or water conditioning agents are compounds that reduce the damage caused by alkaline hydrolysis and adjust the pH of the spray solution so that it is maintained within a pH range of 4 to 6. These compounds are much easier and safer to use than trying to lower the pH of the spray solution with materials such as sulfuric acid. In addition, other materials, such as vinegar (i.e., acetic acid), are often used to acidify water.

You can visit these two Web sites for additional information on the effect of pH on pesticides including insecticides and fungicides:

- http://floriculture.osu.edu/archive/apr04/SpraySolutionPH.html
- http://www.griffins.com/tech_service/bulletins_2003_4_optimum_pesticide_performance.asp.

Chapter 3

Mode of Action and Rotation

Greenhouse producers, in order to sustain successful pest management programs and preserve the longevity of currently available insecticides and miticides need to practice rotating insecticides or miticides to reduce the likelihood that plant-feeding insects and mites in greenhouses will develop resistance. First of all, we must ask the question: What is resistance? Resistance is the genetic ability of some individuals in a pest population to survive an insecticide or miticide application, or a genetic modification that results in the diminished sensitivity of an insect or mite population to a particular insecticide or miticide. In other words, the product no longer effectively kills the target insect or mite pests. This is primarily due to the intensive pressure placed on insect or mite populations from too frequent—or too many—applications of insecticides or miticides with the same mode of action. This results in the amplification of already existing genetic traits and an abundance of resistant individuals and a low number of susceptible individuals within a population. Due to the selection of individuals in insect and mite pest populations to overcome this burden, these populations are then able to tolerate applications of insecticides or miticides. The rate at which insect and mite pests may develop resistance to insecticides or miticides is influenced by a variety of factors, including:

- Length of exposure to a single insecticide or miticide
- Level of mortality (high versus low)
- Presence or absence of refuge sites or hiding places
- Relatedness of a insecticide or miticide to another one
- Generation time (short versus long)
- Number of young or offspring produced per generation

14 Plant Protection

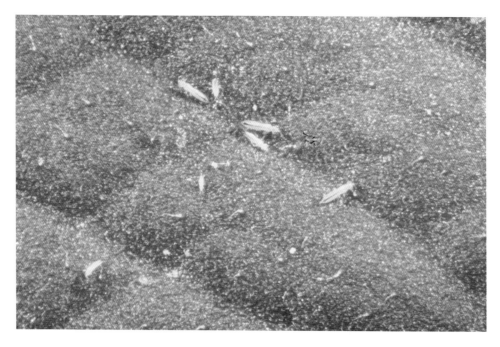

Figure 3.1
Western flower thrips (*Frankliniella occidentalis*), such as the adults shown here, feed on a wide range of ornamental plant species.

◎ Mobility of individuals.

Insect and mite pests may develop resistance to any insecticide or miticide utilizing a number of mechanisms either individually or in combination with one another, including metabolic, physical, physiological, behavioral, or natural.

◎ **Metabolic resistance** is the fragmentation of the active ingredient by an insect or mite pest. When the insecticide or miticide enters the body, enzymes that are present detoxify or convert the active ingredient into a non-toxic form. It is then usually excreted out with other waste products.

◎ **Physical resistance** is an alteration in the cuticle, or skin, that decreases penetration of the insecticide or miticide. For example, young mealybug crawlers do not have a protective covering, which is why they are more

Figure 3.2

Western flower thrips (*F. occidentalis*) adults do not fly very well but may be moved around in greenhouses via air currents.

susceptible to insecticide applications, whereas mature mealybugs tend to possess a white, waxy covering, which inhibits the ability of insecticides to penetrate into the body.

- **Physiological resistance** occurs when an insect or mite pest alters the target site of an insecticide or miticide in such a way that it decreases sensitivity to the active ingredient at the physical point of attachment.
- **Behavioral resistance** occurs when insect or mite pests avoid contact with an insecticide or miticide by residing in locations, such as terminal growing points, leaf or petiole junctures, and the base of plant stems, which may be difficult for an insecticide or miticide to penetrate.

16 Plant Protection

Figure 3.3
Geraniums (*Pelargonium* spp.) grown year round in commercial greenhouse facilities provide a continuous food supply for pests.

◎ **Natural resistance** is a general type of resistance in which the insect or mite pest, or life stage is not susceptible to an insecticide or miticide. In general, the egg and pupa stages of most insect and mite pests are tolerant or not affected by contact insecticides and miticides or systemic insecticides.

Some researchers believe that insect and mite populations will develop resistance faster to insecticides or miticides that are used in greenhouses than when similar products are used outdoors. Their hypothesis is that the material will "hang around" or persist longer in greenhouses than outdoors, whereas insecticides and miticides used outdoors will generally break down from exposure to ultraviolet light (i.e., photodecomposition) or rainfall, which may reduce the amount of selection pressure placed on an insect or mite pest population.

Figure 3.4

Herbaceous and woody plants also provide sustenance for insect and mite pests and may require insecticide or miticide applications.

Factors that may influence the rate of resistance developing within an insect or mite pest population include general operational procedures, insect and mite biological characteristics, and greenhouse production conditions. General operational procedures that may affect the rate of resistance developing include performing insecticide or miticide applications on a frequent basis regardless of insect or mite pest population dynamics, which can be determined by implementing a scouting program. This increases the selection pressure on insect and mite pest populations, thus leading to an increase in resistant individuals and a reduction in susceptible individuals. Consistently applying the highest label rate, continually utilizing the same insecticide or miticide, or using insecticides or miticides with similar modes of activity for an extended period of time also increases the selection pressure placed on insect and mite pest populations.

Figure 3.5

Greenhouse facilities located adjacent to field crops, such as corn or soybean, may experience insects migrating into greenhouses as field crops age or are harvested. This is especially likely when greenhouse openings, such as sidewalls and vents, are not screened.

Biological characteristics of insects and mites that may increase the rate of resistance include rapid development time (i.e., short generation time and rapid transfer of resistant genes), high reproductive rate (i.e., a large number of offspring are produced per generation), and high mobility and wide host range (figures 3.1 and 3.2). It is important to note that resistant genes in insect or mite pest populations can be passed on to future generations or progeny. In fact, all of these biological characteristics may result in increased exposure to insecticide or miticide applications. Additionally, genetic factors influence the rate of resistance. For example, the females of certain insect and mite pests, such as thrips, spider mites, and whiteflies, may be haploid, which means that they possess only a single (i.e., maternal) complement of chromosomes as opposed to having a full complement of maternal and paternal chro-

Figure 3.6

Insecticides and miticides with different modes of action should be incorporated into rotation programs for managing insect and mite pests in greenhouses.

mosomes (i.e., diploid). Females that are haploid tend to concentrate resistant genes. Also, if a particular trait required for resistance occurs through the expression of a single gene (i.e., monogenic resistance) then resistance to an insecticide or miticide may result only after several generations. In contrast, if multiple or numerous genes are required for resistance to develop (i.e., polygenic resistance) then development may occur at a slower rate.

Greenhouse conditions that may lead to an increase in resistant individuals within a population include environmental parameters such as temperature and relative humidity, which are typically conducive for insect and mite development. The greenhouse generally encloses insect and mite pests, which restricts the movement of susceptible individuals into existing populations. Therefore, resistant individuals are dominant and remain in the greenhouse and breed, whereas susceptible individuals from areas not treated with insecticides or miticides are unable to enter and breed with resistant insect and mite pests. Furthermore, intensive year-round production in greenhouses provides a continuous supply of food for insect and mite pests (figures 3.3 and 3.4), which often results in frequent exposure to insecticide or miticide applications.

Resistance may also occur as a result of the movement of insect and/or mite pests within and into greenhouses. In general, there are three ways in which immigration of insect or mite pests may lead to resistance. First, insects or mites that migrate from other crops within a greenhouse or between greenhouses increase the probability that these insect or mite pest populations will be exposed to additional insecticide or miticide applications. Second, receiving plants already infested with insect or mite pests that have been previously exposed to insecticide or miticide applications may enhance the development of resistance because a high percentage of these insect or mite pests may already possess resistant genes. Finally, insect or mite pests that enter a greenhouse from adjacent field or vegetable crops (figure 3.5) may have been exposed to agricultural insecticides or miticides that are very similar to those registered and used in greenhouses.

The rate of resistance development in an insect or mite pest population may vary depending on the season. For example, the number of applications, based on the population dynamics of the insect or mite pest population, can vary throughout the year especially during fall/winter and spring/summer. The probability of resistance developing may increase during spring and summer because this is when insect or mite pest populations are typically higher, thus leading to more frequent applications of either insecticides or miticides.

A basic understanding of the concepts of resistance and the conditions that may enhance the ability of insect and mite pests to develop resistance to insecticides or miticides is important in order to sustain proper insecticide and miticide stewardship programs. The primary management strategy that greenhouse producers can implement, which will potentially minimize problems associated with resistance, is to rotate insecticides and miticides with different modes of action.

The mode of action, or mode of activity, refers to how an insecticide or miticide affects the metabolic and/or physiological processes in an insect or mite pest. In order to reduce the possibility of insect and mite pests developing resistance it is important to design a rotation program that uses insecticides and miticides with different modes of activity—not chemical classes (figure 3.6). The reason for this is that some chemical classes have similar modes of activity. For example, organophosphates and carbamates, despite being different chemical classes, both have identical modes of action. These chemical classes block the action of cholinesterace (ChE), an enzyme that deactivates acetylcholine (ACh) from the nerves, causing nerve signals to continue "firing," resulting in an accumulation of ACh, an exhaustion of energy, and eventually death. So, using acephate, an organophosphate, for two spray applications during a generation and then switching to methiocarb, a carbamate, does not constitute a proper rotation scheme. Similarly, although pyridaben and fenpyroximate are in different chemical classes—pyridazinone and phenoxypyrazole, respectively—they both work on the mitochondria electron transport system (these insecticides/miticides are often referred to a mitochondria electron transport inhibitors or METIs), which is responsible for energy production, so these active ingredients should not be used in succession. The chemical class neonicotinoid contains a number of systemic insecticides that are registered for use in commercial greenhouses. Because all neonicotinoids have similar modes of activity (refer to table 3.1, No. 4), it is important to avoid using them in succession as this will increase the selection pressure on the target insect pest population and may potentially enhance the development of insecticide resistance. Use an insecticide with a different mode of activity either before or after using a neonicotinoid-based insecticide.

Another important strategy is to incorporate insecticides and miticides with non-specific or broad modes of activity, such as insect growth regulators, insecticidal soap, feeding inhibitors or blockers, horticultural oil, and beneficial fungi and bacteria into rotation programs with insecticides and miticides that have specific modes of activity. (Most of the insecticides and miticides discussed above have specific modes of action—that is they are only active on one target site.) Incorporating insecticides and miticides with non-specific or broad modes of activity will reduce the potential of resistance developing. However, it is also important to rotate insect growth regulators with different modes of action (refer to table 3.1, Nos. 7, 13, and 14).

Additionally, it is important to rotate common names (i.e., active ingredient) not trade names. In general, rotate different modes of activity every two to three weeks, or after two to three insect or mite generations. However, this

TABLE 3.1.

Mode of action of insecticides and miticides used in greenhouse production systems*

1. **Acetylcholine Esterase Inhibitors** (1A, 1B)
   Acetylcholine esterase inhibitors inhibit the enzyme cholinesterase (ChE) from clearing the acetylcholine (ACh) transmitter. This prevents the termination of nerve impulse transmission and results in an accumulation of acetylcholine leading to hyperactivity, respiratory failure, exhaustion of metabolic energy, and death.
   Pest control materials:
   - acephate (Orthene/Precise)
   - chlorpyrifos (DuraGuard)
   - methiocarb (Mesurol)

2. **GABA-Gated Chloride Channel Blockers** (2A)
   GABA-gated chloride channel blockers act on the gamma-aminobutyric acid (GABA) receptor by binding to the chloride channels thus preventing chloride ions from entering neurons. This disrupts GABA activity, which leads to hyperexcitation, paralysis, and death.
   Pest control material:
   - endosulfan (Thiodan)

3. **Sodium Channel Blockers** (3)
   Sodium channel blockers destabilize nerve cell membranes by working on the sodium (Na^+) channels in the peripheral and central nervous system and slowing down or preventing closure. This results in stimulating nerve cells to produce repetitive discharges, eventually leading to paralysis and death.
   Pest control materials:
   - bifenthrin (Talstar/Attain)
   - fenpropathrin (Tame)
   - lambda-cyhalothrin (Scimitar)
   - cyfluthrin (Decathlon)
   - fluvalinate (Mavrik)
   - permethrin (Astro)

4. Nicotinic Acetylcholine Receptor Disruptors (4A)

Nicotinic acetylcholine receptor disruptors act on the central nervous system, causing irreversible blockage of the postsynaptic nicotinergic acetylcholine receptors leading to disruption of the nerve transmission and uncontrolled firing of nerves. This results in rapid pulses from a steady influx of sodium (Na^+), leading to hyperexitation, convulsions, paralysis, and death.

Pest control materials:
- acetamiprid (TriStar)
- clothianidin (Celero)
- dinotefuran (Safari)
- imidacloprid (Marathon)
- thiamethoxam (Flagship)

5. Nicotinic Acetylcholine Receptor Agonists (5)

Nicotinic acetylcholine receptor agonists disrupt binding of acetylcholine at nicotinic acetylcholine receptors located at the postsynaptic cell junctures, and negatively affect the gamma-amino butyric acid (GABA)–gated ion channels.

Pest control material:
- spinosad (Conserve)

6. GABA Chloride Channel Activators (6)

GABA chloride channel activators affect gamma-amino butyric acid (GABA)–dependent chloride ion (Cl^-) channels by increasing membrane permeability to chloride ions leading to inhibition of nerve transmission, paralysis, and death.

Pest control material:
- abamectin (Avid)

7. Juvenile Hormone Mimics (7A, 7B, 7C)

Juvenile hormone mimics arrest development by causing insects to remain in a young or immature stage primarily by inhibiting metamorphosis, or change in form. As a result, insects are unable to complete their life cycle.

Pest control materials:
- fenoxycarb (Preclude)
- kinoprene (Enstar II)
- pyriproxyfen (Distance)

TABLE 3.1. continued

8. **Selective Feeding Blockers** (9B, 9C)

   Selective feeding blockers inhibit the feeding behavior of insects by interfering with neural regulation of fluid intake into the mouthparts.

   Pest control materials:
   - flonicamid (Aria)
   - pymetrozine (Endeavor)

9. **Growth and Embryogenesis Inhibitors** (10A)

   Growth and embryogenesis inhibitors disrupt the formation of the embryo during development or inhibit larval maturation. However, the specific mode of action and target site of activity are still unknown.

   Pest control materials:
   - clofentezine (Ovation)
   - hexythiazox (Hexygon)

10. **Disruptors of Insect Mid-gut Membranes** (11A1, 11B2)

    Disruptors of insect mid-gut membranes bind to specific receptor sites on the gut epithelium resulting in degradation of the gut lining and eventual starvation of the insect. Crystals release protein toxins (endotoxins) that bind to the mid-gut membrane receptor sites, creating pores or channels. This paralyzes the digestive system and ruptures the mid-gut cell walls allowing ions to flow through the pores, disrupting potassium (K^+) and pH balances. As a result, the alkaline contents of the gut spill into the blood, resulting in gut paralysis and death.

    Pest control materials:
    - *Bacillus thuringiensis* spp. *israelensis* (Gnatrol)
    - *Bacillus thuringiensis* spp. *kurstaki* (Dipel)

11. **Oxidative Phosphorylation Inhibitors** (12B)

    These inhibit oxidative phosphorylation at the site of dinitrophenol uncoupling, which disrupts the formation or synthesis of adenosine tri-phosphate (ATP).

    Pest control material:
    - fenbutatin-oxide (ProMite)

12. Oxidative Phosphorylation Uncouplers (13)

These uncouple oxidative phosphorylation, which is a major energy-producing step in cells, by disrupting the H^+ gradient, thus preventing the formation of adenosine tri-phosphate (ATP).

Pest control material:
- chlorfenapyr (Pylon)

13. Chitin Synthesis Inhibitors (15, 16, 17)

Chitin synthesis inhibitors prevent the formation of chitin, which is an essential component of an insect's exoskeleton. This causes the insect's cuticle to become thin and brittle. As a result, insects (and mites in the case of etoxazole) die while attempting to molt from one stage to the next.

Pest control materials:
- buprofezin (Talus)
- cyromazine (Citation)
- diflubenzuron (Adept)
- etoxazole (TetraSan)
- novaluron (Pedestal)

14. Ecdysone Antagonists (18, 26)

Ecdysone antagonists disrupt the molting process by inhibiting biosynthesis or metabolism of the molting hormone—ecdysone.

Pest control materials:
- tebufenozide (Confirm)
- azadirachtin (Azatin/Ornazin)**

15. Mitochondria Electron Transport Inhibitors (21, 24)

Mitochondria electron transport inhibitors disrupt Complex 1 electron transport or act on the NADH-CoQ reductase site, or bind to the Qo center of Complex III in the mitochondria, reducing energy production by preventing the synthesis of adenosine tri-phosphate (ATP).

Pest control materials:
- acequinocyl (Shuttle)
- fenpyroximate (Akari)
- pyridaben (Sanmite)

TABLE 3.1. continued

16. Lipid Biosynthesis Inhibitors (23)
Lipid biosynthesis inhibitors block the production of lipids, which are a group of compounds made up of carbon and hydrogen including fatty acids, oils, and waxes. They disrupt the cell membrane integrity and reduce sources of energy.
Pest control material:
- spiromesifen (Judo)

17. GABA-Gated Antagonists (25)
GABA-gated antagonists block or close gamma-aminobutyric acid (GABA) activated chloride (Cl^-) channels in the peripheral nervous system.
Pest control material:
- bifenazate (Floramite)

18. Desiccation or Membrane Disruptors
Desiccation or membrane disruptors damage the waxy layer of the exoskeleton of soft-bodied insects and mites by altering the chitin so that it cannot hold fluids resulting in desiccation (drying up) or smothering insects by covering the breathing pores (spiracles).
Pest control materials:
- neem oil (Triact)
- paraffinic oil (UltraFine Oil)
- potassium salts of fatty acids (insecticidal soap)

---

* Values in parentheses are the mode of action designations based on the Insecticide Resistance Action Committee [IRAC].

** In addition to acting as an insect growth regulator, azadirachtin acts a feeding deterrent/inhibitor, oviposition inhibitor, repellent, egg-laying deterrent, sterilant, and/or direct toxin.

depends upon the time of year, because temperature and season influence the duration of the life cycle. For example, high temperatures that typically occur in greenhouses during the summer months shorten the developmental time (egg to adult) of most major greenhouse insect and mite pests, including aphids, thrips, twospotted spider mites (*T. urticae*), and whiteflies. This often leads to overlapping generations with variable age structures (eggs, larvae, pupae, and/or adults) present at the same time. As a result, more frequent applications of insecticides or miticides are needed, and they must be rotated more often. In contrast, during the winter months, the developmental time of most greenhouse insect and mite pests is extended (due to the cooler temperatures and shorter daylengths), which means that insecticides and miticides may not need to be applied or rotated as frequently.

Below are examples of rotation schemes for aphids, thrips, twospotted spider mites, whiteflies, mealybugs, and fungus gnats using insecticides or miticides (active ingredients are listed) that have dissimilar modes of activity:

- **Aphids:**
  Pymetrozine→Imidacloprid→Paraffinic oil→Acephate→Potassium salts of fatty acids
- **Thrips:** Spinosad→Acephate→Abamectin→Methiocarb
- **Twospotted spider mites:**
  Bifenazate→Chlorfenapyr→Etoxazole→Pyridaben→Spiromesifen
- **Whiteflies:** Dinotefuran→Pyriproxyfen→Bifenthrin→Acephate→Spiromesifen
- **Mealybugs:** Acetamiprid→Acephate→Potassium salts of fatty acids→Kinoprene
- **Fungus gnats:** Pyriproxyfen→Cyromazine→Kinoprene→Chlorpyrifos

Table 3.1 presents a listing of the major modes of activity for insecticides and miticides used in greenhouses with detailed descriptions for each along with the insecticides and miticides (common name and trade name) that may be categorized under each specific mode of action. In addition, the Insecticide Resistance Action Committee (IRAC) mode of action designations are included. These are what appear on the insecticide or miticide label, which will make it easier for greenhouse producers to develop effective rotation and resistance management programs and allow them to deal with insect and mite pests more effectively.

Chapter 4
Tank Mixing

Tank mixing, which involves combining two or more insecticides or miticides into a single spray solution, is a way to improve control of insect and mite pests. The primary reason that greenhouse producers tank mix is convenience. It is less time consuming, costly, and labor intensive to mix two or more insecticides or miticides together into a single solution and perform one spray application as opposed to making two or more applications separately. In addition, tank mixing two or more products together may improve pest control or enhance efficacy. This is referred to as synergism. Synergism occurs when two insecticides and/or miticides are mixed together to perform better than when they are applied separately.

For example, it has been shown that insecticides containing the active ingredient azadirachtin or the entomopathogenic fungus *Beauveria bassiana* are more effective when mixed together or with another insecticide than if either is used by itself. This also has been demonstrated with insecticides and miticides containing paraffinic oil and potassium salts of fatty acids as active ingredients. The reason why these tank mixes are synergistic is primarily because one of the materials somehow stresses the insect enough, thus allowing the other material to work better. For example, during the warmer periods of the year, insects tend to molt (shed old cuticle or skin) so rapidly that the spores of the insect-killing fungus *B. bassiana* cannot penetrate the insect cuticle and start an infestation. However, the addition of an insect growth regulator may stress or slow development enough so that the fungal spores are able to penetrate the insect skin, internally enter the insect's body, and eventually kill it.

Some compounds, often referred too as synergists, are added to certain insecticides and miticides to enhance activity. For example, piperonyl butoxide (PBO), which is not an insecticide, is typically mixed with pyrethroid-based insecticides in order to block enzymes present in insects that are capable of breaking apart or partitioning the active ingredient. Another method that may be used to enhance insecticide activity is to mix a pyrethroid-based insecticide with a different type of insecticide or miticide. The pyrethroid is supposed to, in theory, irritate insect or mite pests causing them to be more active, thus increasing their exposure to spray residues.

Just as synergism improves the performance of two or more insecticides or miticides, the opposite, which is referred to as antagonism may result. Antagonism occurs when mixing two or more insecticides or miticides leads to reduced overall efficacy or mortality compared to if the materials are applied separately. In addition to reduced effectiveness, there is the potential for plant injury or phytotoxicity (see chapter 6). It is important to read the label prior to tank mixing any insecticides or miticides as product labels generally state which materials, such as fungicides, fertilizers, or adjuvants, can and cannot be mixed. If there are any doubts, contact the manufacturer directly for information.

Similarly, incompatibility is a physical condition that prevents some insecticides and miticides from mixing together properly in a spray solution. This may result in reduced efficacy or phytotoxicity. Incompatibility may be due to the chemical or physical nature of the insecticide/miticide, impurities in the water, improper water temperature, or the types of formulations that are mixed. In order to determine if there is compatibility between two or more insecticides or miticides, a "jar test" should be conducted. To conduct a jar test, collect a sample of the spray solution into an empty glass jar or other container and allow the solution to stand for approximately fifteen minutes. If the materials are not compatible, there may be a noticeable separation or layering, or precipitates such as flakes or crystals may form. However, if the insecticides or miticides are compatible, then the solution may appear homogeneous or resemble milk. Many product labels contain information related to compatibility.

It is also important to understand that certain chemical classes have very similar modes of action. For example, both the organophosphates and carbamates, despite being different chemical classes, have identical modes of action (cholinesterace inhibitors). Greenhouse producers should avoid tank mixing insecticides or miticides with similar modes of action (see chapter 3), and insecticides or miticides with site-specific (single target site) modes of action. Tank mixing insecticides or miticides with site-specific modes of action with another one having a broad mode of action, such as insect growth regulators, insecticidal soap, horticultural oil, neem extract, feeding inhibitors or blockers, or beneficial fungi, may minimize the potential of insect or mite populations developing resistance.

Due to the sensitivity of certain plant types and cultivars, tank mixing may lead to phytotoxicity. It is essential that all tank mixes be evaluated for phytotoxicity before any application due to the additive effects of inert ingredients such as carriers, surfactants, and wetting agents contained in certain formulations. Greenhouse producers should avoid tank mixing any products that contain surfactants, because an increase in the surfactant concentration may be phytotoxic to the crop. To avoid phytotoxicity, always test the mix solution on a small sample of plants prior to making an application to an entire crop.

The loss of broad-spectrum insecticides and miticides has resulted in the introduction of alternative insecticides and miticides that have a narrow spectrum of pest activity. Although the availability of products that demonstrate specificity is desirable, it does create a problem when greenhouse producers are dealing with multiple pests. For example, a greenhouse producer that has western flower thrips (*F. occidentalis*), whiteflies, and twospotted spider mite (*T. urticae*) as major pests may have to mix together two (possibly more) insecticides or miticides in order to obtain the same spectrum of control that a single broad-spectrum product may have provided. This may be a concern due to the potential for phytotoxicity and reduced efficacy.

Chapter 5

Coverage and Timing of Application

Thorough, uniform coverage is essential for controlling many greenhouse insect and mite pests (figure 5.1). Greenhouse producers should determine the location of specific insect and mite pest populations by scouting, then direct spray applications to those plant parts to obtain maximum coverage, which will increase the effectiveness of any insecticide or miticide being used. Many insecticides and miticides registered for use in greenhouses have contact activity, which means it is important that sprays are directed toward leaf undersides where a majority of the life stages (eggs, young, and adults) of insect and mite pests, such as twospotted spider mite (*T. urticae*), greenhouse whitefly (*Trialeurodes vaporariorum*), and sweetpotato whitefly (*Bemisia tabaci*), are usually located. However, a number of insecticides and miticides have translaminar, or local systemic, activity. These materials are able to penetrate leaf tissues and form a reservoir of active ingredient within the leaf. This provides residual activity against piercing-sucking insects and mites. The longevity of the residual activity may vary depending on the characteristics of the active ingredient, plant type, and age of plant tissue. A thorough understanding of the biology of insect and mite pests and where certain life stages are located will increase the efficacy of any insecticide and miticide application, whether it be a foliar spray or drench.

Insecticides and miticides should be applied in the early morning or late afternoon because this is when most insects and mites are active. If insecticide or miticide applications are conducted when insect or mite pests are less active, then the level of control may be reduced—particularly for contact materials. Insecticides or miticides applied during hot, dry, sunny days may result in rapid drying and less residual activity—reducing their efficacy and potentially causing plant injury. Applying horticultural oil during cloudy weather may result in phy-

Figure 5.1

Thorough spray coverage, particularly leaf undersides, is important in effectively controlling insect and mite pests of greenhouses.

totoxicity because the material remains on plant leaves for extended periods of time. Evening applications of insecticides and miticides may promote disease development (e.g., *Botrytis cinerea*). In order to reduce the time that plant parts remain moist—heat, vent, and use horizontal airflow fans accordingly throughout the greenhouse.

Certain factors concerning the applicator performing the spray application should also be considered. To avoid inadequate spray coverage, initiate spray applications when the spray applicator is not overly tired. Applications should also be avoided during the heat of the day, when temperature discomfort could lead to reduced spray coverage. In addition, due to the personal protective clothing and equipment required, spraying during the hottest part of the day could lead to possible heat exhaustion of the applicator.

Coverage and Timing of Application **35**

Figure 5.2
A high-volume spray may be required in order to thoroughly cover all plant parts, especially large plants that have an extensive plant architecture, such as this chrysanthemum (*Dendranthema grandiflora*).

Plant size can also impact spray coverage, since it is easier and less time consuming to obtain sufficient coverage on small plants compared to large plants that have a more extensive plant architecture (figure 5.2), thus requiring additional spray time in order to get adequate coverage of all plant parts.

If insecticide or miticide applications are performed when insect or mite pest populations are excessive, then it will take longer to reduce the numbers, and more frequent applications will be required especially when dealing with overlapping generations. In addition, insect and mite pests may have already developed into resistant stages (i.e., pupa), may already be causing plant injury, or may be in locations that are difficult to reach with sprays, such as unopened flower buds. It is more appropriate to make applications when insect and mite pest numbers are low.

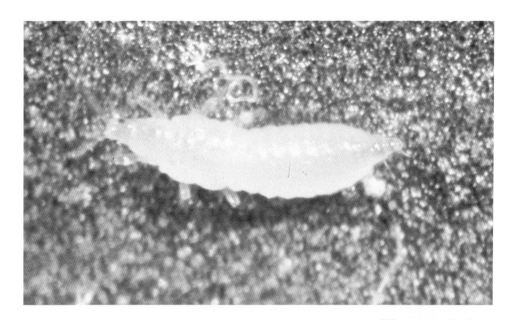

Figure 5.3

Nymphal stage of western flower thrips (*Frankliniella occidentalis*)

Figure 5.4

Immature stage of citrus mealybug (*Planococcus citri*)

Coverage and Timing of Application **37**

Figure 5.5
Twospotted spider mite (*Tetranychus urticae*) egg

Figure 5.6
Greenhouse whitefly (*Trialeurodes vaporariorum*) pupa

Figure 5.7

Yellow sticky cards are used to monitor winged insects, such as the adult stage of thrips, whiteflies, leafminers, fungus gnats, and shore flies.

Insecticide or miticide applications will generally fail if the vulnerable life stage or stages of the target insect or mite pests are not present. The immature or young stages are typically more susceptible than the older life stages to insecticide and miticide applications (figures 5.3 and 5.4). In fact, most contact and systemic insecticides and miticides have minimal effect on the egg and/or pupal stage of many insect and mite pests (figures 5.5 and 5.6). For example, mediocre control will occur if western flower thrips eggs and pupae are the predominant life stages present during an application. The young nymphs that emerge from the eggs, and the adults that emerge from the pupae will not be exposed to an insecticide applied several days previous, especially with short residual insecticides—resulting in the need for additional sprays. This is also important when there are many overlapping generations. In some

Figure 5.8

The presence of particular insect life stages on yellow sticky cards, such as adults, is helpful in timing insecticide applications accordingly.

cases, two spray applications per week may be required in order to decrease the insect or mite pest population below damaging levels before extending the length of the application intervals. Targeting the early developmental stages may reduce the number of insecticide or miticide applications required, thus decreasing the amount of selection pressure placed on an insect or mite pest population, which will reduce the potential for resistance.

Frequency of application may depend on the season. During cooler temperatures, the insect or mite pest life cycle—from egg to adult—is, in general, extended in contrast to warmer temperatures. This may influence the number of applications needed to treat the susceptible life stages. A common problem that occurs in greenhouse production systems is that spray intervals are too long, which often leads to inadequate control. However, certain insecticides

Figure 5.9

Visual inspections of the crop must be performed to assess the presence of spider mites, mealybugs, and scales on plants. In addition, visual plant inspections are helpful in detecting eggs, immatures, and pupae of certain insect and mite pests, such as whiteflies, leafminers, and twospotted spider mite (*Tetranychus urticae*).

and miticides, such as insecticidal soaps and horticultural oils, when applied at frequent intervals may result in phytotoxicity to a crop. Therefore, a proper balance must be found and maintained.

Proper scouting using either colored sticky cards or visual inspections (figures 5.7 through 5.9)—depending on the insect or mite pest—can help detect when the vulnerable life stage is present, and indicate how many applications may be needed to reduce the pest population level. Then an insecticide or miticide can be applied accordingly. Again, understanding the biology of insect and mite pests makes it easier to control populations before they build up to damaging levels.

Chapter 6

Plant Phytotoxicity

Phytotoxicity or plant injury may be the result of plant sensitivity, use of excessive application rates (above those recommended by the insecticide or miticide label), too frequent applications, improper insecticide or miticide dilutions, improper tank mixes (refer to chapter 4), stage of plant growth (flowering versus non-flowering) during application, or environmental conditions. The degree of phytotoxicity may vary depending on the insecticide or miticide formulation (emulsifiable concentrate versus wettable powder), weather conditions (sunny versus cloudy days), ambient air temperature (high versus low), or leaf temperature. Environmental conditions, including temperature and relative humidity, may also have a significant influence on the potential for phytotoxicity after or during application.

The interactions between the insecticides or miticides that occur in mixtures can be a cause of phytotoxicity as well. In addition, plants stressed due to a lack of moisture or nutrients are more susceptible to phytotoxicity from insecticide and miticide applications. It is also important to understand that phytotoxicity is not always associated with the active ingredient of an insecticide or miticide, but may be due to the inert ingredients, such as carriers, solvents, or surfactants in the formulation, or impurities, such as salts in the water, that are introduced during the mixing process.

Symptoms of phytotoxicity include leaf drop, leaf yellowing, stunting, abnormal growth (leaf distortion or leaf curling), necrotic spotting on plant parts, and growth retardation (figures 6.1 through 6.7). Insecticides or miticides may retard plant growth either acutely (i.e., short-term) or chronically (i.e., long term) by affecting plant height or stem elongation. Greenhouse producers that are unsure about the properties of insecticides or miticides should

Figure 6.1

An application of methiocarb (Mesurol) caused the plant injury to this African daisy (*Osteospermum ecklonis*).

test them on a small quantity of plants or contact the manufacturer. This is less costly than discovering later, when your entire crop is unmarketable. In general, check plants seven to ten days after the application to evaluate any injury. However, for some insecticides and miticides you will need to wait fourteen to twenty-eight days for phytotoxicity symptoms to be expressed. Be sure to record any problems for future reference. In most cases, insecticides or miticides will not cause any plant injury when used according to label rates—however, if the label rate is increased two, three, or four times what is recommended, then the potential for phytotoxicity exists. Although tank mixing may be advantageous, there are phytotoxic risks. If two or more insecticides or miticides are not compatible, this can result in significant injury to a crop. Greenhouse producers should refer to the label for information on the com-

Figure 6.2

Acephate (Orthene) spray injury

patibility of insecticides or miticides in mixtures and perform a sample test of all mixtures before treating an entire crop.

In order to avoid phytotoxicity, it is essential that spray equipment be cleaned thoroughly between applications. Also, never use one sprayer for herbicides, insecticides, and fungicides—always have a dedicated sprayer for herbicides as well as one dedicated for insecticides and fungicides. Be sure that the spray mixture receives frequent agitation during the application in order to prevent settling in the spray tank, which can increase the possibility for phytotoxicity as the solution grows stronger the more the tank empties.

Manufacturer labels typically list the plants on which a given insecticide or miticide is safe to use. However, it is difficult—if not impossible—to evaluate all products registered for use in greenhouses for phytotoxicity to the wide

Figure 6.3

The leaf necrosis shown here is an example of insecticide spray injury.

diversity of plant cultivars grown. Additionally, certain product labels will state that the insecticide or miticide should not be applied to plants that are in flower or bract, as these plant parts are more sensitive than leaves.

Insecticidal soaps and horticultural oils, when applied at frequent intervals (for example, three times in one week) may damage sensitive plants. Horticultural oils can harm the new growth and foliage of particularly sensitive plant species. Problems associated with horticultural oils may be due to high humidity (less than 70 percent), overcast and cloudy days, inadequate airflow, high temperature and/or light intensity, and too frequent applications. All of these conditions may result in phytotoxicity.

The labels of insecticides and miticides typically contain restrictions or precautionary statements that provide information on any potential phytotoxici-

Figure 6.4

Insecticidal soap (potassium salts of fatty acids) applied too frequently can cause injury to plants, such as this impatiens (*Impatiens walleriana*).

ty issues. Below are examples from randomly selected insecticide and miticide labels that either have restrictions or precautionary statements related to phytotoxicity:

- "Phytotoxicity has been observed following application of this product on geraniums (*Pelargonium* spp.) and *Impatiens* spp. It is therefore recommended that this product not be used on geraniums or impatiens."
- "This product has been evaluated for phytotoxicity on a wide range of crops. However, since all combinations or sequences of pesticide sprays, including fertilizers, surfactants, and adjuvants, have not all been tested, it is recommended that a small area be sprayed first to make certain that no phytotoxicity occurs."

Figure 6.5

The insecticide spray injury on these asters (*Aster* spp.) features a uniform pattern of symptoms

- "To ensure that this product is compatible with the variety or cultivar under your specific conditions, test the product on a limited scale and observe for phytotoxicity for two weeks before making large-scale applications. Phytotoxicity has been observed on the following plants: salvia (*Salvia* spp.), ghost plant (*Graptopetalum paraguayense*), Boston fern (*Nephrolepis exaltata*), schefflera (*Schefflera* spp.), gardenia (*Gardenia* spp.), and coral bells (*Heuchera sanguinea*). It is therefore recommended that this product not be used on these plants."
- "This product has been evaluated for phytotoxicity on a wide range of ornamental plants. However, since all combinations or sequences of pesticide sprays, including surfactants and adjuvants, have not been tested, it is recommended that a small area be sprayed first to make certain that no

Figure 6.6

An improper drench application of the insect growth regulator pyriproxyfen (Distance) caused the injury to this poinsettia (*Euphorbia pulcherrima*).

phytotoxicity occurs. Phytotoxicity has been observed following the use of this product on certain species of ferns (e.g., *Adiantum* spp.) and Shasta daisy (*Leucanthemum* spp.). It is therefore recommended that this product not be used on ferns or Shasta daisy."

- "Certain pansy cultivars have exhibited sensitivity to this product. Prior to use on pansy, apply this product at the appropriate use rate to a small portion of the pansy crop (about 10 units per cultivar) and visually assess impact seven to ten days after application."

Phytotoxicity can be a serious problem in reducing the market value of a crop or even make the crop unsaleable—both cases translate into an economic loss. Greenhouse producers can avoid phytotoxicity problems by: 1) read-

Figure 6.7

The injury to this poinsettia (*Euphorbia pulcherrima*) displays the characteristic upward leaf cupping more commonly from a phenoxy-based herbicide, such as 2,4-D; however an improper drench application of the insect growth regulator pyriproxyfen (Distance) caused the injury.

ing the label of all insecticides and miticides prior to application to make sure that they are registered for use on a particular plant species; 2) using insecticides and miticides at the recommended label rate and the appropriate intervals of application; 3) testing all new insecticides or miticides on a small number of plants before using on an entire crop; 4) making sure all spray equipment is well maintained and properly calibrated; 5) frequently agitating the spray solution during an application; 6) avoiding applications during high temperatures (more than 85° F [29° C]) and when plants are exposed to intense sunlight; and 7) avoiding application to water- or nutrient-stressed plants.

Chapter 7
Pesticide Storage

Proper storage of insecticides and miticides is an important indirect component of pest management, as it extends their shelf life, preserving their longevity and effectiveness. Insecticides and miticides do not carry lifetime warranties, so you must use them within a specified time period. Always store insecticides and miticides in their original containers, and protect them from moisture and temperature extremes. Store insecticides and miticides in a separate building or area away from people (i.e., employees and customers) and food. Avoid storing clothes, respirators, or any personal items in the same area as insecticides and miticides. Make sure the storage area is well ventilated and dry and has adequate lighting. Lock all doors and windows when not in use. Be watchful for unauthorized persons in the storage facility and application equipment areas. Also, be sure to report any suspicious activity to the proper authorities (i.e., police). You must post clear, visible warning signs near all primary entryways to let personnel know that the building or area contains insecticides and miticides. Maintain records of all insecticides and miticides being stored. Records should include the date of purchase and the date each insecticide or miticide was placed in storage. This is important because in general, insecticide and miticide labels do not state when the product was produced or manufactured. However, be sure to note the batch number on the label in case you have any questions or problems. Keep chemical records in a separate building as well as with the local fire department so that in the case of a fire emergency workers can be notified what materials are inside.

Insulated storage chambers are ideal for protecting insecticides and miticides from environmental conditions (figure 7.1). In general, the ideal storage conditions for most insecticides and miticides are temperatures between 55

50 Plant Protection

Figure 7.1

Proper storage is essential in maintaining the shelf life of insecticides and miticides.

and 65° F (13 and 18° C) and a relative humidity between 40 and 50 percent. Check the storage area regularly to ensure that containers are in good condition and there are no leaks or spills present. Any spill must be cleaned up immediately and disposed of properly. Avoid storing herbicides next to insecticides or miticides, as vapors from certain herbicides, such as 2,4-D, may combine with insecticides and miticides and result in cross-contamination and potential damage to treated plants after an application. Be sure that all containers are tightly sealed.

Some insecticides and miticides do not store well for extended periods of time. These materials have a short shelf life. For example, liquid formulations, if not used after a certain period of time (over two years), may eventually separate into layers, settle out, and form precipitates in the bottom of

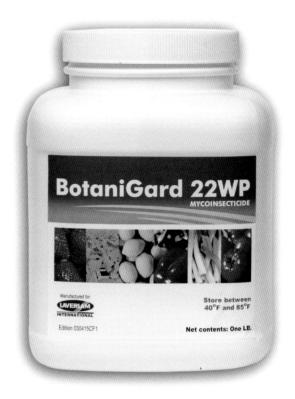

Figure 7.2

Botanigard contains the beneficial fungus *Beauveria bassiana*.

containers. This makes it difficult to get the active ingredient or carrier back into a suspension so that the insecticide or miticide is suitable for use. It is recommended to shake insecticides or miticides formulated as liquids or solutions prior to use in order to obtain a homogeneous mixture. In addition, prolonged storage, especially after temperature extremes, may cause chemical changes, resulting in materials losing their effectiveness or increasing their potential to harm plants. In general, most insecticides and miticides should not be stored any longer than three to four years; however, this depends on the formulation, storage conditions, and inert and active ingredients. For example, products containing living organisms, such as fungi (figure 7.2), bacteria (figure 7.3), or entomopathogenic nematodes (figure 7.4), may have a shorter shelf life than conventional materials. Additionally, solvents and

Figure 7.3

Gnatrol contains the beneficial bacterium *Bacillus thuringiensis* spp. *israelensis*.

petroleum-based materials may deteriorate certain types of containers after a period of time.

Environmental conditions, including high temperatures, ultraviolet light, and high humidity, may cause insecticides or miticides to break down or degrade. Many insecticides and miticides break down when exposed to extreme heat (over 100° F [37° C]) or cold (below 32° F [0° C]) over an extended period of time. Insecticides and miticides formulated as liquids expand when heated or frozen, which may cause containers to rupture. Proper ventilation of the storage area is important to prevent overheating as high temperatures can reduce the efficacy of certain insecticides or miticides. Low temperatures may cause some materials to settle or crystallize out of solution. The relative humidity during storage may alter the composition of

Figure 7.4

Nemasys contains the entomopathogenic nematode *Steinernema feltiae*.

some insecticides and miticides, especially those stored in unsealed containers. For example, solid formulations such as dusts, wettable powders, and water-dispersible granules may cake when exposed to high relative humidity. Be sure to keep containers or packages on shelves or pallets to reduce exposure to excess moisture.

Chapter 8

Identification of Insect and Mite Pests

Aphids

Aphids can be major insect pests of greenhouses, particularly early in the season, feeding on many types of bedding and potted plants. There are a number of different aphid species that attack greenhouse-grown plants including the green peach aphid (*Myzus persicae*), melon or cotton aphid (*Aphis gossypii*), foxglove aphid (*Aulacorthum solani*), chrysanthemum aphid (*Macrosiphoniella sanborni*), oleander aphid (*Aphis nerii*), and potato aphid (*Macrosiphum euphorbiae*) (figs. 8.1. and 8.2). Aphid color will vary depending on the particular host plant fed upon and as such should not be used for identification. Certain plant types and cultivars are more susceptible to aphids than others.

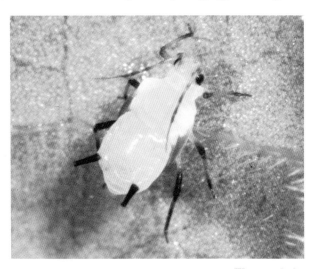

Figure 8.1
Oleander aphid (*Aphis nerii*)

Aphids, in general, are approximately 1 to 2 mm long, soft-bodied insects that possess tubes (i.e., cornicles) on the end of their abdomens. Females do not need to mate to reproduce; through a process known as parthenogenesis, females give birth to live female offspring that can themselves start producing their own young or nymphs within seven to ten days. Each female can give birth to up to one hundred live young per day for a period of twenty to thirty days. The rapid reproductive capacity of aphids results in the potential for large populations developing within a short period of time.

Figure 8.2
Close up of aphid feeding

Aphid reproduction, however, may be influenced by plant quality and nutrition.

Aphids typically feed on new terminal growth and on leaf undersides (figs. 8.3 and 8.4). However, distribution varies depending on the plant species or cultivar. Aphids cause direct plant injury by removing plant fluids with their piercing-sucking mouthparts. Feeding on new growth results in young leaves appearing distorted or curled (upward or downward). They may also cause plant stunting. Aphids also produce a clear, sticky liquid material referred to as honeydew, which is a growing medium for black, sooty mold fungi. The presence of sooty mold detracts from the aesthetic quality of plants. High aphid populations may lead to the presence of white cast

Figure 8.3
Oleander aphids feed on a variety of plants species.

or molting skins that can also reduce plant aesthetics. In addition, many aphids are capable of transmitting destructive viruses.

Aphid control involves implementing cultural, insecticidal, and biological control strategies, preferably all three. Proper watering and fertility practices are effective in minimizing potential problems with aphids. Avoid overfertilizing plants, as aphids are attracted to and feed on plants receiving excess amounts of nitrogen. Aphid reproduction increases when they feed on plants that have been overfertilized with nitrogen-based fertilizers. Sanitation is an important means of avoiding problems with aphids. Remove plant debris and old stock plants, from the greenhouse or place these materials into containers with tight-sealing lids, because winged aphids will abandon desiccating plant material and migrate onto the main crop. Weed removal will eliminate potential aphid hosts, since many broadleaf and grassy weed species commonly found in and around greenhouses serve as a reservoir for aphids and can support large populations.

Figure 8.4
Aphids feed on the undersides of leaves.

Aphids are susceptible to contact, translaminar, and systemic insecticides. Contact insecticides, including insect growth regulators, insecticidal soap, horticultural oil, and pyrethroid-based insecticides, are effective against aphids. However, more than one application is usually required in order to successfully reduce aphid populations. Contact insecticides are generally most effective early in the crop cycle, because the smaller plant size makes it easier for sprays to penetrate the crop canopy, ensuring adequate coverage of leaf undersides. Insecticides with translaminar activity, meaning that the material penetrates and resides in the leaf tissues forming a reservoir of active ingredient, may provide residual activity, up to fourteen days in some cases, even after spray residues have dried. However, the length of effective residual activity is dependent on the insecticide. Systemics, such as the neonicotinoid-based insecticides and selective-feeding blockers, are also effective against aphids, especially when applied early in the crop cycle and before aphid populations build up. Systemic insecticides may be applied as a drench to the growing medium or to plant leaves. There are a number of insecticides that have both translaminar and systemic properties. When applying systemic insecticides as either a drench or granule, it is essential to make applications to every plant container because those plants that fail to receive any insecticide may serve as a reservoir for aphids. Be sure to rotate insecticides with different modes of action in order to avoid resistant aphid populations from developing.

Biological control is another strategy that may be successful in dealing with aphids. However, it is important to identify the aphid species because various parasitoids are specific to the aphid species they will attack. Releases of either parasitoids or predators must be made before aphid populations build up and plant damage is evident. The biological control agents (i.e., natural enemies) commercially available for aphids include the parasitoids *Aphidius colemani, A. matricariae*, and *A. ervi*; the predatory midge *Aphidoletes aphidimyza*; the green lacewings *Chrysoperla carnea* and *C. rufilabris*; the ladybird beetle *Hippodamia convergens*; and the beneficial fungus *Beauveria bassiana* (sold as BotaniGard and Naturalis-O).

Broad and Cyclamen Mites

Broad mite (*Polyphagotarsonemus latus*) and cyclamen mite (*Steneotarsonemus pallidus*) are pests of certain greenhouse-grown crops such as begonia, cyclamen, fuchsia, Transvaal daisy, and impatiens (figs. 8.5 and 8.6). Broad and cyclamen mites require very different environmental conditions than twospotted spider mite (discussed on page 71). These mites tend to be a problem under cool temperatures (around 59° F [15° C]) and high relative humidity (70 to 80%), which are conducive for their development and reproduction. Both mites have similar developmental and reproductive potential. Broad and cyclamen mites feed on young foliage and floral parts, such as flower buds, retarding growth and preventing flowers from fully developing. Typical symptoms of feeding by broad and cyclamen mites include bronzing and distorted plant leaves. Excessive populations often lead to these mites feeding on both the upper and lower portions of the entire leaf surface. The presence of either broad or cyclamen mites is detected typically after plant injury is noticeable rather than actually finding the mites themselves.

Broad mite adults are approximately 0.25 mm long, amber to dark green in color, and oval in shape with a white strip extending down the back (fig. 8.7). Eggs are oval shaped and covered with bumps. Broad mites tend to feed on the underside of young leaves. The life cycle from egg to adult takes less than one week to complete, with females capable of laying up to twenty-five eggs. Cyclamen mite adults are also 0.25 mm long; however, their eggs are oval and smooth. Females are yellow-brown in color. Their life cycle, from egg to adult, may be completed in one to three weeks depending on temperature. Cyclamen mite females can lay one to three eggs per day; potentially laying up to sixteen eggs. Symptoms of cyclamen mite feeding include bronzing and curling of leaves. Heavy infestations may cause leaves to appear brittle and turn brown to silver. In addition, distorted or twisted leaves are reduced in size.

Figure 8.5
Broad mite (*Polyphagotarsonemus latus*) damage to Transvaal daisy (*Gerbera jamesonii*). Note the purplish, distorted leaves.

Figure 8.6
Broad mite damage on English ivy (*Hedera helix*)

Miticides with translaminar properties are best for control of broad and cyclamen mites, since these mites are normally protected from contact miticides. Preventative applications are required because once damage is evident it may be too late to control the infestation. The removal of plants exhibiting symptoms as well as those surrounding the symptomatic plants is essential in order to prevent the spread of broad and cyclamen mites onto other plants. The predatory mite *Neoseiulus barkeri* may be used to control both broad and cyclamen mites.

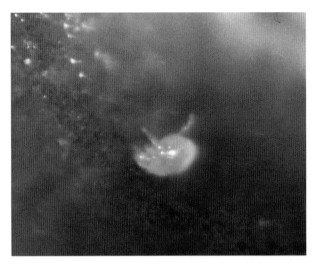

Figure 8.7
Broad mite adult (*Polyphagotarsonemus latus*)

Fungus Gnats

Fungus gnats (*Bradysia* spp.) may be a problem in greenhouses for several reasons. First, large populations of adults flying around may affect crop salability. Second, both the adult and larval stages are capable of disseminating and transmitting diseases. Third, larvae feed on roots causing direct plant injury and create wounds that allow soilborne pathogens to enter. Finally, larvae may tunnel into the crown of a plant and cause plant death (figs. 8.8 and 8.9).

Fungus gnat adults are winged, 4.0 mm long insects with long legs and antennae (fig. 8.10). They tend to fly around the growing medium surface and live for approximately seven to ten days. Females can deposit up to two hundred eggs into cracks and crevices located in the growing medium. Eggs hatch into transparent or slightly translucent, legless larvae that are approximately 6.0 mm in length. A characteristic feature of fungus gnat larvae is the presence of a black head capsule (fig. 8.11). Larvae are generally located within the top 2.5 to 5.0 cm of the growing medium. However, they can also be found throughout the growing medium and in the bottom of containers near drainage holes.

The life cycle, from egg to adult, can be completed in twenty to twenty-eight days, depending on ambient air and growing medium temperatures.

Proper sanitation, such as the removal of weeds, old plant material, and old growing medium, as well as the elimination of algae, will reduce potential problems associated with fungus gnat populations. Weeds growing underneath benches create a moist environment that is conducive for fungus gnat development. Hand pulling or using an herbicide will kill existing weeds. Avoid overwatering and overfertilizing plants, as this leads to conditions that promote algae growth. Keep floors, benches, and cooling pads free of algae by using a disinfectant, such as one containing quaternary ammonium salts.

Insecticides are commonly used to control fungus gnats; however, they must be used in conjunction with algae control. Do not rely solely on insecticides to control fungus gnats. Many of the insecticides registered for fungus gnats are insect growth regulators. Insect growth regulators are only effec-

Figure 8.8
Shown here is damage to a Transvaal daisy (*Gerbera jamesonii*) crop due to fungus gnats.

Figure 8.9
Fungus gnat larvae root feeding injury on *Pelargonium*

tive on the larval stage, as they have no direct adult activity. Several conventional and microbial-based insecticides may be used against fungus gnats. These insecticides are applied as drenches or sprenches into containers, or they are applied directly to gravel or soil floors in order to kill fungus gnat larvae. Adult fungus gnats may be controlled using conventional insecticide sprays or aerosols.

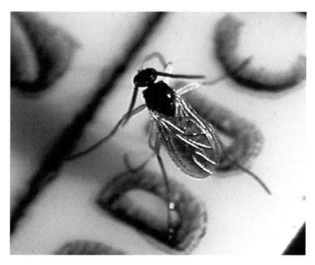

Figure 8.10
Adult fungus gnat on yellow sticky card

Biological control is another option to manage fungus gnats in greenhouse production systems. Biological control agents or natural enemies that are effective in controlling fungus gnats are the beneficial nematode, *Steinernema feltiae*; the soil-predatory mite, *Hypoaspis miles*; and the rove beetle, *Atheta coriaria*. All three biological control agents attack fungus gnat larvae. They are applied to the growing medium or soil floor and must be applied early before the population builds up to damaging levels.

Figure 8.11
Close up of fungus gnat larvae

Leafminers

The serpentine leafminer (*Liriomyza trifolii*) is typically the main species that attacks horticultural greenhouse crops, although other species may also appear depending on geographic location. Leafminer larvae cause the primary plant damage since they feed between the leaf surfaces in the mesophyll layer of cells, creating serpentine mines or trails.

The damage caused by leafminers is generally aesthetic, and plants are rarely killed from an infestation. However, a heavy infestation may impact plant salability, particularly potted plants (fig. 8.12). An adult female leafminer may also cause damage when puncturing leaves with her ovipositor to lay eggs. This creates white specks on the leaf surfaces. Females may puncture both the upper and lower leaf surface; however, this depends on the species. Leaf puncturing may reduce photosynthesis and kill young plants.

Figure 8.12
Leafminer larvae feeding damage on chrysanthemum (*Dendranthema* × *grandiflorum*)

Leafminer adults are small, 2 to 3.5 mm long, shiny, black flies with yellow markings on their abdomen. Females live for two to three weeks, laying an average of sixty eggs during their lifespan. Eggs hatch into bright yellow to white larvae (or maggots) that feed on the mesophyll layer of cells creating mines within the leaf (fig. 8.13). The larvae may either mine the top of the leaf, the bottom of the leaf, or both, depending on the species.

Mines enlarge in size as the larva grows or molts to the next instar. The last larval instar cuts a semicircular slit in the leaf and drops to the growing medium or soil to pupate. Pupae are oblong and brown to gold in color. Leafminers require darkness in order to pupate, so they are typically located deep within the growing medium or soil profile. Adults emerge from the pupal stage in approximately ten days. The life cycle, from egg to adult, is completed in sixteen to twenty-four days, depending on ambient air and growing medium temperatures.

Implementing cultural control strategies is important in minimizing problems with leafminers. Avoid overfertilizing plants, especially with nitrogen-based fertilizers, since plants receiving excessive nitrogen levels are more susceptible to attack. Weed and plant debris removal inside greenhouses and outdoors will eliminate alternative hosts for leafminers.

Figure 8.13
Leafminer larvae feeding damage on garden verbena (*Verbena* × *hybrida*) leaf

Screening greenhouse openings, such as vents and sidewalls, is extremely helpful in preventing adult leafminers (as well as western flower thrips and whiteflies) from migrating into a greenhouse from outside. Removing leaves from infested plants or removing highly infested plants before leafminers pupate will reduce numbers. Placing yellow sticky tape (similar to CAUTION tape) around a susceptible crop and/or underneath benches of greenhouses with gravel or soil floors will capture adult leafminers, and in the long-term, possibly reduce populations (fig. 8.14).

Insecticides may be used to manage leafminer populations; however, several species have developed resistance to commonly used insecticides, which has complicated leafminer control. Additionally, larvae are well protected within the leaf tissue, thus escaping contact from insecticide applications. Insect growth regulators can be used to target the larval stage. Insecticides with translaminar properties are effective against the larvae because these materials are capable of entering the leaf and killing the larvae directly. Pyrethroid-based insecticides are useful against adults; however, these materials are generally not effective on the larvae. The number of leafminers present and the occurrence of overlapping generations may influence the frequency of insecticide applications. Sprays should be applied in the morning when females are laying eggs in order to disrupt this behavior. Certain neonicotinoid-based insecticides have activity on several species.

Figure 8.14
Leafminer adults can be captured on yellow sticky tape or yellow sticky cards.

Biological control of leafminers primarily involves the use of parasitoids. Parasitoids in the genus *Diglyphus*, including *D. isaea, D. intermedius, D. begini,* may be used to control certain leafminers. *Diglyphus* spp. are more effective during the warmer temperatures of the spring and summer, while *Dacnusa sibirica* works best during the cooler temperatures of the winter and early spring. Parasitoids are attracted to yellow sticky cards, so cards and tape should be removed prior to release.

Mealybugs

Mealybugs may be a problem on many greenhouse-grown crops, mainly because they are difficult to manage with most insecticides. The major mealybug species found in greenhouses are the citrus mealybug, *Planococcus citri* (fig. 8.15), and in some instances the long-tailed mealybug, *Pseudococcus longispinus*. The long-tailed mealybug, however, tends to be more of a problem in conservatories. In addition to these aboveground mealybugs, there are mealybugs that feed on roots (*Rhizoecus* spp.), which may make control difficult with currently available insecticides. There are also a number of introduced mealybugs, such as the pink hibiscus mealybug (*Maconellicoccus hirsutus*) and the Maderia mealybug (*Phenacoccus madeirensis*). Both species have a wide host range, which includes many types of horticultural plants grown both in greenhouses and outdoors.

Mealybugs cause direct plant injury by feeding on plant fluids with their piercing-sucking mouthparts. This causes leaf yellowing, plant wilting, and stunting (fig. 8.16). In addition, mealybugs excrete a clear, sticky liquid called honeydew, which serves as an excellent growing medium for black, sooty mold fungi. They prefer to congregate in large numbers at leaf junctures where the petiole meets the stem, on the underside of leaves, on stem tips, and under leaf sheaths of certain plant types.

Figure 8.15
Citrus mealybugs feeding on coleus
(*Solenostemon scutellarioides*)

Figure 8.16
Mealybugs feeding on chrysanthemum
(*Dendranthema* × *grandiflorum*)

The mealybug life cycle consists of an egg stage, several immature (i.e., crawler) stages (fig. 8.17), and an adult stage. Before females die, they lay eggs underneath their bodies (with the exception of long-tailed mealybug females that give birth to live offspring). The eggs hatch into mobile crawlers that move around on plants, seeking a place to settle and feed. Once settled, mealybugs go through several growth stages (i.e., instars) before becoming adults . Males eventually become winged individuals that mate with females and then die, whereas females continue development and then die after laying her full complement of eggs. The eggs remain protected under the body of the dead female until they hatch. A single female is capable of laying up to six hundred eggs (fig. 8.18). The life cycle, from egg to adult, takes approximately sixty days, depending on ambient air temperatures.

Figure 8.17
Shown here are the immature stages of citrus mealybug, *Planococcus citri*.

Figure 8.18
Citrus mealybug female typically lays eggs within a cottony mass.

Since females do not fly, they will not be captured on yellow sticky cards, thus visual plant inspections will be required in order to detect early mealybug infestations. Adults are difficult to control because they form a waxy, protective covering that is impervious to most insecticides. In addition, since most insecticides have no activity on eggs, at least two to three applications within a week may be necessary to achieve satisfactory control, especially when dealing with overlapping generations. Crawlers do not possess a waxy covering and are susceptible to insecticides including insect growth regulators, insecticidal soap, horticultural oil, and insect-killing fungi.

Figure 8.19
Mealybugs are commonly found feeding along the leaf midrib or vein of many plants.

Thorough coverage is essential when using contact insecticides because mealybugs are located in inaccessible areas such as the base of leaf petioles, leaf sheaths, and the underside of leaves (fig. 8.19). The addition of a spreader-sticker to the spray solution may help in improving coverage and penetration; however, be sure to read the label for any restrictions in regards to adding spreader-stickers. Systemic insecticides, such as the neonicotenoids, are effective on the feeding stages of mealybugs. However, they need to be applied before populations build up and while the plants are actively growing, so the active ingredient is translocated to areas where mealybugs feed.

The use of biological control agents (i.e., natural enemies) may be successful in controlling mealybugs under certain crop production systems. The natural enemies currently available for control of the citrus mealybug include the predaceous ladybird beetle *Cryptolaemus montrouzieri*, also referred to as the mealybug destroyer, and the parasitoid *Leptomastix dactylopii*. Both are very effective in controlling the citrus mealybug and can be used together.

Shoreflies

Shoreflies, *Scatella* spp., are considered greenhouse insect pests and are generally a problem under moist conditions, especially during propagation and plug production. However, they can be a problem year round. In general, shoreflies are a nuisance pest because their presence, especially in large numbers, can detract from the plant's overall appearance. Adult shoreflies are also capable of transmitting certain soilborne diseases.

Shoreflies have a life cycle consisting of an egg, three larval, a pupal, and an adult stage. A generation can be completed in fifteen to twenty days, depending on ambient air and growing medium temperatures. Adults resemble houseflies (fig. 8.20). They are approximately 3.1 mm long and deep black in color. Each forewing usually has at least five light-colored spots. The antennae and legs are short and the head is small. Larvae are 6.3 mm long and opaque to yellowish-brown in color with no black head capsule. Shorefly adults are stronger fliers than fungus gnat adults.

Figure 8.20
Shorefly adults on young plants are a nuisance pest.

Shoreflies are less likely to cause direct plant damage compared to fungus gnats. However, they are a concern because they are more noticeable flying around plants and are easily seen, especially when captured on yellow sticky cards. Large numbers may be moved from one greenhouse to another on plant material. Shipping plants with large numbers of shorefly adults flying around may reduce crop salability. Although adults are generally considered a nuisance pest, they can leave black fecal deposits on plant leaves that may affect the plant's aesthetic quality (fig. 8.21). Larvae primarily feed on algae located on the surface of the growing medium. The larvae can also be found within the growing medium, although they do not normally feed on plant roots. Adults and larvae are capable of transmitting black root rot, *Thielaviopsis basicola*, in their frass. In addition, adults have been shown to vector the water-mold fungus, *Pythium aphanidermatum*. Adult frass deposited on the lower leaves and stems of susceptible plants may lead to host infection if proper environmental conditions are present.

Proper sanitation such as removing weeds, old plant material, and old growing medium can reduce shorefly problems. Weeds growing underneath benches create a moist environment that is conducive for development. Hand pulling or using an herbicide is effective in killing existing weeds. Most importantly, eliminate algae build-up. Avoid overwatering and overfertilizing plants as this leads to conditions that promote algae growth. Keep floors, benches, and cooling pads free of algae by using a disinfectant or a material containing quaternary ammonium salts. Greenhouse growers may want to consider placing yellow sticky tape (similar to CAUTION tape) underneath greenhouse benches or above the crop canopy in order to capture shorefly adults.

Figure 8.21
Shorefly fecal deposits (i.e., the black specks shown here) on plant leaves can reduce the aesthetics of the plant and affect salability.

Insecticides may be used to control shoreflies; however, they must be used in conjunction with algae control. Do not rely solely on insecticides to control shoreflies. Many of the insecticides registered for shoreflies are insect growth regulators. Insect growth regulators are targeted toward the larval stage and have no direct adult activity. The bacterial insecticide, *Bacillus thuringiensis* subsp. *israelensis*, which is used to control fungus gnat larvae has no activity on shorefly larvae. Insecticides can be applied as drenches or sprenches into containers or applied directly to gravel or soil floors. Adults may be controlled with insecticide sprays or aerosols. However, because shoreflies may be located throughout the greenhouse, widespread distribution is essential.

The predatory mite *Hypoaspis miles* may be used to manage shoreflies. It can be applied directly to the soil beneath greenhouse benches. However, biological control of shoreflies is generally not effective because they live and develop under moist conditions that are not conducive for survival of most natural enemies.

Twospotted Spider Mite

Twospotted spider mite (*Tetranychus urticae*) feeds on a wide range of crops grown in greenhouses, primarily from spring through early fall (fig. 8.22). They feed on leaf undersides within plant cells with their stylet-like mouthparts, damaging the spongy mesophyll, palisade parenchyma, and chloroplasts, thus reducing chlorophyll content and the plant's ability to photosynthesize. Damaged leaves appear stippled with small, silvery-gray to yellowish speckles. Heavily infested leaves may appear bronzed, turn brown, and fall off. The mites may also spin irregular webbing, which allows them to move among plants—especially when plants are spaced close together and leaves are in contact with each other (fig. 8.23). Twospotted spider mites may also be carried on wind currents or via crop handling. These mites prefer warm, dry conditions with low relative humidity.

The twospotted spider mite is approximately 1.6 mm long and oval shaped. It can vary in color from greenish-yellow to reddish-orange (fig 8.24). Adult females possess distinct black spots located on

Figure 8.22
Twospotted spider mite feeding injury on sword lily (*Gladiolus* spp.)

both sides of their body. The females live about thirty days and can lay up to two hundred small, spherical, transparent eggs on leaf undersides. The life cycle from egg to adult takes one to two weeks to complete, depending on ambient air temperatures. For example, the life cycle from egg to adult takes fourteen days at 70° F (21° C) and seven days at 84° F (29° C).

Figure 8.23
Shown here is a heavy infestation of twospotted spider mites on New Guinea impatiens. Note the webbing, which allows the mites to disperse among plants.

Twospotted spider mite management involves combining cultural practices with miticide usage. Cultural practices that will reduce problems include: 1) avoid overfertilizing plants, especially with soluble forms of nitrogen, as this results in the production of soft, succulent tissue that is easier for twospotted spider mite to penetrate with their mouthparts; 2) remove old plant material, which may serve as an inoculum source of mite populations when the next crop is initiated; 3) avoid water-stressing plants, because this increases susceptibility to mites; and 4) remove weeds from within and around greenhouses, since weeds such as nightshades and creeping woodsorrel serve as mite hosts.

There are numerous miticides or insecticide/miticides available for controlling twospotted spider mite populations. Most have contact activity, so thorough coverage of all plant parts, especially the underside of leaves, is essential. A number of materials have translaminar activity. Be careful when using insecticidal soap or horticultural oil to control twospotted spider mites, as repeat applications may be phytotoxic to plants. In addition, read the label to determine which life stages (i.e., egg, larva, nymph, or adult) each material works best on.

It is extremely important to rotate miticides or insecticide/miticides with different modes of action in order to reduce the possibility of resistance. In general, greenhouse producers should only use a material once or twice within a generation and then switch to another material with a different mode of action.

Biological control or the use of natural enemies is another viable management strategy. There are a number of predatory mites commercially available, which if used properly, can provide effective control of twospotted spider mite populations. The predatory mite most widely used is *Phytoseiulus persimilis*. However, there are other predatory mites that may be more appropriate depending on the environmental conditions. It is important to note that predatory mites must be released before twospotted spider mite populations reach damaging levels.

Figure 8.24
Adult twospotted spider mite (*Tetranychus urticae*)

Western Flower Thrips

Western flower thrips (*Frankliniella occidentalis*) feeds on a wide range of greenhouse-grown crops (figs. 8.25 through 8.28). Their feeding on plant leaves may result in leaf scarring, necrotic spotting, distorted growth, and sunken tissues (primarily on leaf undersides). In addition, adults feeding on flowers or unopened buds may lead to flower bud abortion or deformed flowers. Flowers and leaves typically have a characteristic white or silvery appearance. Black fecal deposits may be present on leaf undersides.

Western flower thrips are small, slender insects approximately 2.0 mm in length with fringed or hairy wings. They vary in color from yellow-brown to dark brown (fig. 8.29). Adult females insert eggs into leaves, laying up to two hundred fifty eggs during their forty-five-day lifespan. Eggs hatch into nymphs that feed on leaves and flowers. Western flower thrips will pupate in flowers, leaf litter, or growing media. Adults that emerge from the pupal stage typically feed on flowers. The life cycle, from egg to adult, takes approximately three weeks to complete, depending on ambient air temperatures.

Sanitation practices, such as removing weeds, old plant material debris, and growing medium debris, will alleviate problems with western flower thrips. Certain weeds, particularly those with yellow flowers such as common groundsel (*Senecio vulgaris*), may attract adults. Remove plant material debris from the greenhouse or place into containers with tight-sealing lids, since western flower thrips will abandon desiccating plant material and migrate onto the main crop. Screening greenhouse openings, such as vents and sidewalls, will prevent the thrips from entering greenhouses from outside. The appropriate screen size or mesh for western

Figure 8.25
Western flower thrips feeding injury on poinsettia

Figure 8.26
Western flower thrips feeding injury on verbena (*Verbena* × *hybrida*) flowers

flower thrips is 192 microns (i.e., 132-mesh).

The principal management strategy involves the use of insecticides. Initiate applications when populations are low, which avoids having to deal with different age structures or life stages, such as egg, nymph, pupa, and adult, simultaneously over an extended time period. Once the populations reach damaging levels, then more frequent applications are required. Insecticides with contact or translaminar activity are generally used to control western flower thrips because systemic insecticides do not usually move into flower parts (e.g., petals and sepals) where adults typically feed. Insecticides must be applied prior to the thrips' entering terminal or flower buds because once they do, it is very difficult to obtain adequate control and thus prevent injury. As a result, translaminar insecticides are more likely to be effective in killing western

Figure 8.27
Western flower thrips (*Frankliniella occidentalis*) feeding injury on chrysanthemum (*Dendranthema* × *grandiflorum*) leaves

Figure 8.28
Western flower thrips (*Frankliniella occidentalis*) feeding damage on Transvaal daisy (*Gerbera jamesonii*) flower

flower thrips in the terminal or flower buds. However, applications conducted after flowers open are generally too late, since injury has already occurred. In general, high-volume sprays are needed in order to kill thrips that are located in hidden areas of plants, such as unopened flower buds.

Most currently available insecticides only kill the nymphs or adults, with no or minimal activity on either the egg or pupal stages. As such, repeat applications are typically required in order to kill the life stages that were not affected by previous insecticide applications. This is especially important when overlapping generations are prevalent. Three to five applications within a seven- to ten-day period are generally required when populations are high and there are different life stages or overlapping generations present. Frequency of application depends on the time of year. During cooler temperatures the western flower thrips' life cycle is extended compared to warmer temperatures.

Figure 8.29
Adult western flower thrips (*Frankliniella occidentalis*)

The rotation of insecticides with different modes of action is essential to prevent or minimize the potential for resistance. In general, rotate chemicals with different modes of action every two to three weeks or within a single generation. However, this will vary with the time of year, since the development rate of the life cycle is temperature dependent.

Biological control of western flower thrips relies on using natural enemies, such as predatory mites (*Neoseiulus* or *Amblyseius* spp.), minute pirate bugs (*Orius* spp.), and entomopathogenic fungi (*Beauveria bassiana*). However, the key to implementing a successful biological control program is to release natural enemies early enough in the cropping cycle. Releases must be initiated prior to the thrips entering terminal or flower buds. Natural enemies will not control an already established or existing damaging western flower thrips population because it takes time from initial release before natural enemies will lower population numbers below damaging levels.

Whiteflies

The major whitefly species that feed on greenhouse-grown crops include the greenhouse whitefly (*Trialeurodes vaporariorum*) and silverleaf whitefly (*Bemisia argentifolii*), which is synonymous with the sweetpotato whitefly (*Bemisia tabaci*) B-biotype (fig. 8.30). Life stages (i.e., egg, nymph, pupa, and adult) are located on the underside of plant leaves (fig. 8.31). The nymphs cause direct plant injury by feeding on plant fluids, which results in leaf yellowing, leaf distortion (i.e., curling), and plant stunting and wilting. The nymphs also produce a clear, sticky liquid material called honeydew that serves as a growing medium for black, sooty mold fungi. The presence of large numbers of whitefly adults can also be a visual nuisance, which may impact plant salability.

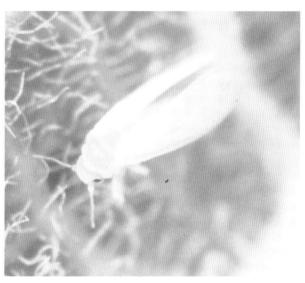

Figure 8.30
Adult sweetpotato whitefly (*Bemisia tabaci* B-biotype)

Adult whiteflies are white, narrow-shaped, and about 2.0 to 3.0 mm in length. Adult females deposit eggs on leaf undersides in a crescent-shaped pattern. Eggs hatch into nymphs or crawlers that migrate short distances then settle down to feed (figs. 8.32, 8.33, and 8.34). The life cycle, from egg to adult, takes approximately thirty-five days; however, this is dependent on ambient air temperatures (fig. 8.35). A single female whitefly can lay eggs one to three days after emerging as an adult.

Each female may live for approximately thirty days and lay up to two hundred eggs.

Whitefly control involves implementing cultural, insecticidal, and biological control strategies, preferably all three (fig. 8.36). Proper watering and fertility practices are essential in minimizing potential problems since whiteflies tend to feed on and lay more eggs on plants that receive abundant levels of nitrogen-based fertilizers. Sanitation is always the most important means of avoiding and/or reducing problems. Remove plant debris and old stock plants from the greenhouse or place into containers with tight-sealing lids because whiteflies will abandon desiccating plant material and migrate onto the main crop. Weed removal in and around the greenhouse will eliminate potential sources of whiteflies, since certain weeds such as sow thistle (*Sonchus* spp.) and creeping woodsorrel (*Oxalis corniculata*) may harbor

Figure 8.31
Whitefly life stages—including egg, nymph, pupa, and adult—are typically found on the underside of leaves, such as this poinsettia.

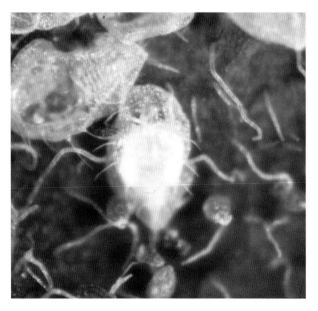

Figure 8.32
Whitefly adult developing inside the pupa

whitefly populations. Screening greenhouse openings such as vents and sidewalls will prevent whiteflies from entering greenhouses from outside. It is recommended to use the same screen size or mesh as for western flower thrips (i.e., 192 microns or 132-mesh) so that both insect pests may be restricted from entering greenhouses from outdoors.

Whiteflies are susceptible to contact, translaminar, and systemic insecticides. Contact insecticides, including many insect growth regulators, insecticidal soap, horticultural oil, and pyrethroid-based insecticides, are generally effective. However, more than one application is usually required since these insecticides are primarily active on two life stages: nymphs and adults. In fact, insect growth regulators only kill whitefly nymphs and have no direct adult activity. In addition, contact insecticides are generally most effective early in the crop cycle because the smaller plant size makes it easier for sprays to penetrate the crop canopy, ensuring adequate coverage of leaf undersides. Translaminar insecticides may also provide control of whiteflies even

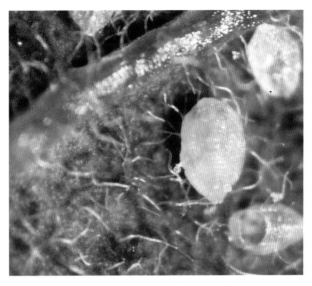
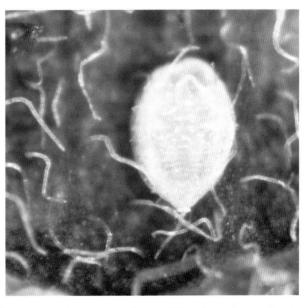

Figures 8.33 and 8.34
Shown are whitefly pupae (top), and close up of a pupa (bottom).

after residues dry. Systemics, such as the neonicotinoid-based insecticides and selective feeding blockers, are effective, especially when applied early in the crop cycle and before whiteflies reach damaging numbers. Systemic insecticides may be applied as a drench to the growing medium or to plant foliage. There are a number of insecticides that have both translaminar and systemic properties.

Biological control is another strategy that may be successful in dealing with whiteflies. This involves using parasitoids, predators, or beneficial (i.e., entomopathogenic) fungi. Commercially available biological control agents (i.e., natural enemies) include the parasitoids *Encarsia formosa*, *Eretmocerus eremicus*, and *E. mundus*; the predatory ladybird beetle *Delphastus catalinae*, and the beneficial fungus *Beauveria bassiana* (sold as BotaniGard and Naturalis-O).

Figure 8.35
Adult whitefly, along with nymph and pupae, are found on the underside of these leaves.

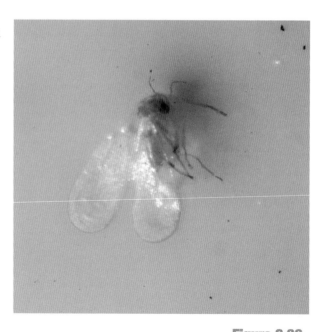

Figure 8.36
Monitor adult whitefly populations using yellow sticky cards.

About the Author

Raymond A. Cloyd is associate professor, extension specialist in ornamental entomology/integrated pest management at Kansas State University. His research and extension program involves pest management in greenhouses, nurseries, landscapes, turfgrass, conservatories, and interiorscapes. Dr. Cloyd has published over thirty scientific refereed publications and over two hundred trade journal articles on topics related to pest management. In addition, he has authored and co-authored books (*Pests and Diseases of Herbaceous Perennials, IPM for Gardeners*), book chapters, manuals, and many extension-related publications. Raymond has received a number of awards including: University of Illinois Extension Outstanding/Innovative Program Team Award ("Insect Identification Series"); Southern Illinois Bedding Plant School Outstanding Service Award; American Society of Horticultural Science, Outstanding Extension Publication Award; Excellence in Extension Campus Award For Less than 10 Years of Service 2004; and Early Career Extension Award from Epsilon Sigma Phi Extension Fraternity. Raymond is a frequent speaker at state, national, and international conferences and seminars.

Selected References

Blanchette, R., ed. 2001. *GrowerTalks on Pest Control*. Batavia, Ill.: Ball Publishing.

Cloyd, R. A. 2003. Managing insects and mites. In *Ball Redbook: Crop Production*, 17th ed., ed. D. Hamrick, 113–125. Batavia, Ill.: Ball Publishing.

Dent, D. 1991. Insecticides. In *Insect Pest Management*, 132–212. Wallingford, U.K.: CAB International.

Dole, J. M., and H. F. Wilkins. 2005. Pest management. In *Floriculture Principles and Species*, 162–183. Upper Saddle River, N.J.: Pearson Education, Inc.

Dreistadt, S. H. 2001. *Integrated Pest Management for Floriculture and Nurseries*. 3402. Oakland, Calif.: University of California, Statewide Integrated Pest Management Project, Division of Agriculture and Natural Resources.

Marer, P. J. 1988. *The Safe and Effective Use of Pesticides*. 3324. Oakland, Calif.: University of California, Statewide Integrated Pest Management Project, Division of Agriculture and Natural Resources.

Pedigo, L. P. 2002. Conventional insecticides for management. In *Entomology and Pest Management*, 381–440. Upper Saddle River, N.J.: Pearson Education, Inc.

Powell, C. C., and R. K. Lindquist. 1997. *Ball Pest & Disease Manual*, 2nd ed. Batavia, Ill.: Ball Publishing.

Stenersen, J. 2004. *Chemical Pesticides: Mode of Action and Toxicology.* Boca Raton, Fla.: CRC Press.

Thomson, W. T. 2001. *Agricultural Chemicals, Book I—Insecticides.* Fresno, Calif.: Thomson Publications.

Ware, G. W., and D. M. Whitacre. 2004. *The Pesticide Book.* Willoughby, Ohio: MeisterPro Information Resources.

Index

*Page numbers with *f* indicate figures;
page numbers with *t* indicate tables.

A

Abnormal growth, 41
Acephate (Orthene), 10
    injury from, 43f
Acetylcholine (ACh), 21
Acetylcholine esterase inhibitors, 22t
Adiantum spp., 47
African daisy (*Osteospermum ecklonis*), 42f
Alkaline hydrolysis, 9, 11
Amino acids, 2, 3
Amylase, 3
Antagonism, 30
Aphids, 3, 4, 55–58
    control of, 57–58
    description of, 55–56
    dietary nitrogen for, 3
    feeding behavior of, 56, 56f, 57f
    piercing-sucking mouthparts in, 1
    reproduction of, 56
    rotation schemes for, 27
Aphis gossypii (melon or cotton aphid), 4, 55
Aphis nerii (oleander aphid), 2f, 55, 55f, 56f
Applications, coverage and timing of, 33–40
Asters (*Aster* spp.), 46f
Aulacorthum solani (foxglove aphid), 55
Azadirachtin, 9, 10

B

Bacillus thuringiensis spp. *israelensis*, 52f
Beauveria bassiana, 29, 51f
Beetles, chewing mouthparts in, 1
Behavioral resistance, 15
Bemisia argentifolii (silverleaf whitefly), 77
Bemisia tabaci (sweetpotato whitefly),
    33, 77, 77f

Bifenazate, 11
Boston fern (*Nephrolepis exaltata*), 46
Botanigard, 51f
Botrytis cinerea, 34
Bradysia spp. (fungus gnats), 6f, 60–62
Broad mites (*Polyphagotasonemus latus*),
    7, 58–60, 59f

C

Carbamates, modes of action of, 21, 30
Carbaryl, 9–10, 10t
Carbohydrases, 3
Caterpillars, 8
    chewing mouthparts in, 1
Chewing insects, 1, 5f, 8
Chitin synthesis inhibitors, 25t
Cholinesterace (ChE), 21
Cholinesterace inhibitors, 30
Chrysanthemum aphid
    (*Macrosiphoniella sanborni*), 55
Chrysanthemum
    (*Dendranthema grandiflorum*), 3f, 4, 35f
    leafminer damage to, 63f
    mealybugs damage to, 66f
    western flower thrips damage to, 73, 75f
Citrus mealybugs (*Planococcus citri*),
    immature stage of, 36f, 67f
Coleus (*Solenostemon scutellarioides*),
    mealybug damage to, 66f
Contact insecticides, 8
Coral bells (*Heuchera sanguinea*), 46
Cotton aphid (*Aphis gossypii*), 4, 55
Cross-contamination, 50
Cyclamen mites
    (*Steneotarsonemus pallidus*), 7, 58–60

85

D

Dendranthema grandiflorum (chrysanthemum), 3f, 4, 35f
Desiccation disruptors, 26t
Disruptors of insect mid-gut membranes, 24t

E

Ecdysone antagonists, 25t
English ivy (*Hedera helix*), mite damage of, 59f
Environmental conditions, 52
Eriophyid mites, 7
Euphorbia pulcherrima (poinsettia), 6f, 47f, 48f

F

Fenpyroximate, 21
Fenvalerate, 10
Flea beetles, 5f
Foxglove aphid (*Aulacorthum solani*), 55
Frankliniella occidentalis (western flower thrips), 7, 14f, 15f, 31, 73–77
Free amino acids, 3
Fungus gnats (*Bradysia* spp.), 6f, 60–62
    chewing mouthparts in, 1
    larvae, 62f
    rotation schemes for, 27
    on yellow sticky card, 62f

G

GABA chloride channel activators, 23t
GABA-gated antagonists, 26t
GABA-gated chloride channel blockers, 22t
Gardenia (*Gardenia* spp.), 46
Garden verbena (*Verbena hybrida*), leafminer damage to, 64f
Genetic factors, influence on rate of resistance, 18–19
Geraniums (*Pelargonium* spp.), 16f, 45
Gerbera jamesonii (Transvaal daisy), 61f
Ghost plant (*Graptopetalum paraguayense*), 46
Graptopetalum paraguayense (ghost plant), 46
Greenhouse whitefly (*Trialeurodes vaporariorum*), 33, 77
    pupa, 37f
Green peach aphid (*Myzus persicae*), 4, 55
Growth and embryogenesis inhibitors, 24t
Growth retardation, 41

H

Hedera helix (English ivy), 59f
Heuchera sanguinea (coral bells), 46
Honeydew, 56
Horticultural oils, 44
    application of, during cloudy weather, 33–34

I

Impatiens (*Impatiens walleriana*), 45
Incompatibility, 30
Insecticidal soaps, 44
Insecticide Resistance Action Committee (IRAC), 27
Insecticides
    effects of pH on miticides and, 9–11
    mode of action of, used in greenhouse production systems, 22–26t
Insulated storage chambers, 49–50, 50f

J

Jar test, 30
Juvenile hormone mimics, 23t

L

Leaf distortion, 1
Leaf drop, 41
Leafminers, 63–65
    larvae, 3f, 7
    on yellow sticky tape or yellow sticky cards, 65f
Leaf necrosis, 44f
Leaf yellowing, 41
Leucanthemum spp. (Shasta daisy), 47
Lewis mites, 7
Lipid biosynthesis inhibitors, 26t

M

Macrosiphum euphorbiae (potato aphid), 55
Macrosiphoniella sanborni (chrysanthemum aphid), 55
Mealybugs, 4, 65–68
    dietary nitrogen for, 3
    feeding by, 66, 66f, 68f
    laying of eggs of, 67f
    piercing-sucking mouthparts in, 1
    rotation schemes for, 27
Melon aphid (*Aphis gossypii*), 4, 55
Membrane disruptors, 26t
Metabolic resistance, 14

Methiocarb, 11
Miticides, 8
    effects of pH on insecticides and, 9–11
    mode of action of, used in greenhouse production systems, 22–26t
Mitochondria electron transport inhibitors (METIs), 21, 25t
Mode of action, 20–21
Monogenic resistance, 19
Myzus persicae (green peach aphid), 4, 55

N

Natural resistance, 15–16
Necrotic spotting, 41
Nemasys, 53f
Neonicotinoid-based insecticides, 6, 8
Neonicotinoids, 7, 21
Neoseiulus barkeri (predatory mite), 60
Nephrolepis exaltata (Boston fern), 46
Nicotinic acetylcholine receptor, 6
Nicotinic acetylcholine receptor agonists, 23t
Nicotinic acetylcholine receptor disruptors, 23t
Nitrogen, 1–2
Nitrogen-based fertilizers, 2

O

Oleander aphid (*Aphis nerii*), 2f, 55, 55f, 56f
Organophosphates, modes of action of, 21, 30
Osteospermum ecklonis (African daisy), 42f
Oxidative phosphorylation inhibitors, 24t
Oxidative phosphorylation uncouplers, 25t

P

Palisade parenchyma, 8
Parthenogenesis, 56
Pelargonium spp. (geraniums), 16f, 45
    fungus gnat damage to, 61f
Pesticide storage, 49–53
    ideal temperatures in, 49–50
    lighting, 49
    prolonged, 51
    record keeping, 49
    shelf life, 50–51, 50f
    ventilation and, 49
pH
    defined, 9
    effects of, on insecticides and miticides, 9–11

Phenoxypyrazole, 21
Phloem, 1
Photodecomposition, 16
Physical resistance, 14–15
Physiological resistance, 15
Phytoseiulus persimilis, 73
Phytotoxicity
    avoiding, 43
    observation of, 46–47
    potential for, 30, 31
    as serious problem, 47–48
    tank mixing and, 31
Piercing-sucking mouthparts, insects with, 1, 2–3, 5–6
Piperonyl butoxide (PBO), 29–30
Planococcus citri diflora (citrus mealybug), 36f, 65–66, 66f, 67f
Plant phytotoxicity, 41–48
    cause of, 41
    degree of, 41
    symptoms of, 41–42
Plant size, impact on spray coverage, 35
Plant wilting, 1
Poinsettia (*Euphorbia pulcherrima*), 6f, 47f, 48f
    western flower thrips damage to, 73, 74f
Polyphagotasonemus latus (broad mites), 7, 58–60, 59f
Postsynaptic nicotinergic acetylcholine receptors, 6
Potato aphid (*Macrosiphum euphorbiae*), 55
Predatory mite (*Neoseiulus barkeri*), 60
Protein, 2, 3
Pyrethroid-based insecticides, 30
Pyridaben, 21
Pyridazinone, 21

R

Records, maintaining, 49
Relative humidity during storage, 52–53
Resistance, 9
    behavioral, 15
    defined, 13
    factors influencing development of, 16–17
    rate of, 13
    genetic factors influence on rate of, 18–19
    metabolic, 14
    monogenic, 19
    natural, 15–16

physical, 14–15
physiological, 15
rate of development, 20

S

Salvia (*Salvia* spp.), 46
Schefflera (*Schefflera* spp.), 46
Season, frequency of application and, 39–40
Secondary metabolites, 1
Selective-feeding blockers, 6, 7, 24*t*
Shasta daisy (*Leucanthemum* spp.), 47
Shorefly, 69–71
    fecal deposits of, 70, 70*f*
    as nuisance pest, 69, 69*f*
Silverleaf whitefly (*Bemisia argentifolii*), 77
Sodium channel blockers, 22*t*
Soft scale, 4
    piercing-sucking mouthparts in, 1
Solenostemon scutellarioides (coleus), 66*f*
Spider mites, 7
Spill, cleanup of, 50
Spray coverage, 34*f*
    impact on plant size, 35
Spray intervals, length of, 39
Steinernema feltiae, 53*f*
Steneotarsonemus pallidus (cyclamen mites), 7, 58–60
Stunting, 1, 41
Sweetpotato whitefly (*Bemisia tabaci*), 33, 77, 77*f*
Synergism, 29–30
Systemic insecticides, 5, 6–7, 8

T

Tank mixing, 29–31
    phytotoxic risks in, 42
Tetranychus urticae (twospotted spider mite), 4*f*, 7–8, 31, 33, 71–73
Thrips, rotation schemes for, 27
Translaminar, or local systemic activity, 33
Transvaal daisy (*Gerbera jamesonii*)
    fungus gnat damage to, 63*f*
    mite damage of, 59*f*
    western flower thrips damage to, 73, 75*f*
Trialeurodes vaporariorum (greenhouse whitefly), 33, 77
True bugs, 4
2,4-D, 50

Twospotted spider mite (*Tetranychus urticae*), 4*f*, 7–8, 31, 33, 71–73
    adult, 73*f*
    description of, 71–72
    egg of, 37*f*
    feeding by, 71, 71*f*
    infestation of, on New Guinea impatiens, 72*f*
    management of, 72–73
    rotation schemes for, 27

U

Uniform coverage, importance of, 33, 34*f*

V

Variegated plants, 3
Verbena (*Verbena hybrida*), 63*f*, 64*f*
    western flower thrips damage to, 73, 74*f*
Visual inspections, 40, 40*f*

W

Weevils, chewing mouthparts in, 1
Western flower thrips (*Frankliniella occidentalis*), 7, 14*f*, 15*f*, 31, 73–77
    description of, 74, 76*f*
    feeding injury from, 74–75*f*
    management strategy of, 74–76
    nymphal stage of, 36*f*
Whiteflies, 4, 31, 77–80
    control of, 78–80
    description of, 77–78
    dietary nitrogen for, 3
    life stages of, 78, 78*f*, 79*f*, 80*f*
    piercing-sucking mouthparts in, 1
    rotation schemes for, 27

X

Xylem, 2, 4

Y

Yellow sticky cards, 40, 65*f*
    in monitoring winged insects, 38*f*, 39*f*